Decision Support System for Diabetes Healthcare: Advancements and Applications

RIVER PUBLISHERS SERIES IN BIOTECHNOLOGY AND MEDICAL RESEARCH

Series Editors:

PAOLO DI NARDO
University of Rome Tor Vergata, Italy

PRANELA RAMESHWAR
Rutgers University, USA

Aiming primarily at providing detailed snapshots of critical issues in biotechnology and medicine that are reaching a tipping point in financial investment or industrial deployment, the scope of the series encompasses various specialty areas including pharmaceutical sciences and healthcare, industrial biotechnology, and biomaterials. Areas of primary interest comprise immunology, virology, microbiology, molecular biology, stem cells, hematopoiesis, oncology, regenerative medicine, biologics, polymer science, formulation and drug delivery, renewable chemicals, manufacturing, and biorefineries.

Each volume presents comprehensive review and opinion articles covering all fundamental aspect of the focus topic. The editors/authors of each volume are experts in their respective fields and publications are peer-reviewed.

For a list of other books in this series, visit www.riverpublishers.com

Decision Support System for Diabetes Healthcare: Advancements and Applications

Editors

Usha Desai
S.E.A College of Engineering & Technology, Bengaluru, India

Biswaranjan Acharya
Marwadi University, India

Madhu Shukla
Marwadi University, India

Varadraj Gurupur
University of Central Florida, USA

NEW YORK AND LONDON

Published 2025 by River Publishers
River Publishers
Alsbjergvej 10, 9260 Gistrup, Denmark
www.riverpublishers.com

Distributed exclusively by Routledge
605 Third Avenue, New York, NY 10017, USA
4 Park Square, Milton Park, Abingdon, Oxon OX14 4RN

NEW YORK AND LONDON

Decision Support System for Diabetes Healthcare: Advancements and Applications / by Usha Desai, Biswaranjan Acharya, Madhu Shukla, Varadraj Gurupur.

© 2025 River Publishers. All rights reserved. No part of this publication may be reproduced, stored in a retrieval systems, or transmitted in any form or by any means, mechanical, photocopying, recording or otherwise, without prior written permission of the publishers.

Routledge is an imprint of the Taylor & Francis Group, an informa business

ISBN 978-87-7004-166-9 (hardback)
ISBN 978-87-7004-702-9 (paperback)
ISBN 978-87-7004-689-3 (online)
ISBN 978-87-7004-688-6 (master ebook)

While every effort is made to provide dependable information, the publisher, authors, and editors cannot be held responsible for any errors or omissions.

Contents

Preface xiii

List of Figures xv

List of Tables xix

List of Contributors xxi

List of Notations and Abbreviations xxv

1 Importance of Analyzing Causality for Diabetes Care 1
Veena Mayya, Christian King, and Varadraj Gurupur
1.1 Prevalence of Diabetes 2
1.2 Factors Contributing to Diabetes 5
1.3 Decision Support Tools and Diabetes Management 6
1.4 Conclusions 9

2 Advances and Opportunities in Digital Diabetic Healthcare Systems 15
Nasir Vadia, Priya Patel, and Vijaykumar Sutariya
2.1 Background 16
2.2 Transformation of Diabetes Management from Conventional to Digital 18
2.3 Digital Technologies for Diabetes Management 20
    2.3.1 Artificial intelligence and machine learning (AI and ML) 21
    2.3.2 Medical and healthcare Internet of Things (MHIoT) 25
    2.3.3 Blockchain 27
    2.3.4 Telemedicine 30
        2.3.4.1 Telephonic/mobile care 31
        2.3.4.2 Systems for continuous glucose monitoring 33

		2.3.4.3	Diabetic retinopathy control techniques	33
		2.3.4.4	Insulin pen	34
		2.3.4.5	Insulin therapy decision support systems	34
		2.3.4.6	Asymptomatic diabetes screening	34
	2.3.5	mHealth		36
2.4	Current Challenges and Future Perspective			41
2.5	Conclusion			45

3 Role of IoT and Expert System in Diabetes Control with Continuous Diagnosis of Medical Conditions   53

*Tarun Kumar Vashishth, Kewal Krishan Sharma,
Vikas Sharma, Sachin Chaudhary, Bhupendra Kumar,
and Rajneesh Panwar*

- 3.1 Introduction 54
- 3.2 History of Diabetes 55
- 3.3 Diabetes 55
  - 3.3.1 Type-1 diabetes 56
  - 3.3.2 Type 2 diabetes 56
  - 3.3.3 Gestational diabetes 57
  - 3.3.4 Pre-Diabetes 57
- 3.4 Progressive Nature of Diabetes 57
- 3.5 Symptoms of Diabetes 58
- 3.6 Individual Diabetes Control Program 59
- 3.7 IoT 60
- 3.8 Expert Systems 62
- 3.9 Literature Review 63
- 3.10 Discussions 64
  - 3.10.1 Process of developing type-2 and type 1 diabetes . . 66
  - 3.10.2 Advantages of IoT use in individual diabetic control 66
  - 3.10.3 Disadvantages of IoT use in individual diabetic control 68
- 3.11 Conclusion 74
- 3.12 Future Scope 76

4 Harnessing Machine Intelligence and Big Data for Diabetes Management   81

Pranjul Mishra, Nancy Jadeja, and Madhu Shukla

- 4.1 Introduction 82
- 4.2 Machine Learning in Diabetic Care 83

	4.2.1	Supervised learning methodologies in diabetic disease diagnosis	84
		4.2.1.1 Logistic regression: A precision tool for binary classification	86
		4.2.1.2 Unveiling precision with support vector machines	86
		4.2.1.3 Carving paths with decision trees in diabetic classification	87
		4.2.1.4 Pioneering precision with neural networks	88
	4.2.2	Unveiling the unsupervised: insights through data elevation	90
		4.2.2.1 Clustering: Carving pathways to personalized diabetic care	90
		4.2.2.2 Anomaly detection: Illuminating unseen risks in diabetic care	92
		4.2.2.3 Dimensionality reduction: Navigating complexity in diabetic insight	93
4.3	The Enigmatic Influence of Big Data in Diabetic Care		94
4.4	Navigating Challenges and Seizing Opportunities in Diabetic Care		95
4.5	Ethical and Privacy Frontiers in Diabetes Management		96
4.6	Unleashing Transformation: Unveiling the Potential of Big Data and Machine Learning		98
4.7	Converging Horizons: Illuminating Emerging Trends and Technologies		99
4.8	Navigating the Uncharted: Forging Ahead in Diabetes Care		100
4.9	Conclusion		101

5 Machine Intelligence and Big Data in Diabetic Care: Laboratorian's Perspective — 107

Arindam Ghosh, Aritri Bir, and Asitava Deb Roy

5.1	Introduction		108
5.2	Machine Intelligence in the Field of Laboratory Science		108
	5.2.1	The significance of data in diabetic care	111
	5.2.2	The role of big data analytics in diabetic care	111
	5.2.3	The role of machine intelligence in diabetic care	111
	5.2.4	The role of laboratory medicine in diabetic care	112
	5.2.5	Personalized diabetes management	112

viii Contents

        5.2.6   Early detection and prevention of diabetes complications 113
        5.2.7   Predictive analytics for diabetes management 113
                5.2.7.1   Results and impact 114
                5.2.7.2   Challenges and considerations 115
        5.2.8   Personalized treatment plans 116
        5.2.9   Remote monitoring and telemedicine 119
        5.2.10 Challenges and opportunities 119
  5.3   Conclusion 120

6 EfficientNetB3-DTL: Classification of Diabetic Retinopathy Images using Modified EfficientNetB3 with Deep Transfer Learning   125

Ch.Rajendra Prasad, Sreedhar Kollem, Srinivas Samala, B. Srinivas, Ravichander Janapati, and Srikanth Yalabaka

  6.1   Introduction 126
  6.2   Related Work 127
  6.3   Proposed Model 129
        6.3.1   Dataset 130
        6.3.2   Data pre-processing 131
        6.3.3   Deep transfer learning 132
  6.4   Results and Discussion 137
  6.5   Conclusion 141

7 Prediction and Diagnosis of Glaucoma in Fundus Images through Optic Cup and Optic Disk Segmentation   145

M. Ponnibala, Usha Desai, Biswaranjan Acharya, Vassilis C. Gerogiannis, and Andreas Kanavos

  7.1   Introduction 146
  7.2   Related Work 148
        7.2.1   Color-based segmentation 149
        7.2.2   Model-based segmentation 149
        7.2.3   Texture-based segmentation 149
  7.3   Material and Methods 151
        7.3.1   Retinal image acquisition 151
        7.3.2   Proposed methodology 151
        7.3.3   Active contour segmentation for optic disc localization 152
        7.3.4   Level set algorithm for optic cup localization 152

		7.3.5	K-means clustering based segmentation algorithm	153
	7.4	Feature Extraction on the Segmented Optic Disk and Cup		155
		7.4.1	Cup to disc ratio (CDR)	155
		7.4.2	Texture-based feature extraction	158
	7.5	Experimental Analysis		159
		7.5.1	Multi-class support vector machine (SVM) classifier	159
		7.5.2	Extreme learning machine (ELM) classifier	161
		7.5.3	Discussion	163
	7.6	Conclusions and Future Work		164

8 Early Diagnosis of Diabetes using an Intelligent Machine Learning Technique   169
C. V. Guru Rao and Nilgün Şengöz

	8.1	Introduction			170
	8.2	Related Works			172
	8.3	System Model and Problem Statement			174
	8.4	Proposed Model			175
		8.4.1	Process of the proposed methodology		176
			8.4.1.1	Data training and preprocessing	176
			8.4.1.2	Feature analysis	177
			8.4.1.3	Classification and prediction	178
			8.4.1.4	Gene expression analysis	178
	8.5	Result and Discussion			181
		8.5.1	Case study		181
		8.5.2	Performance assessment		182
			8.5.2.1	Precision	184
			8.5.2.2	Accuracy	184
			8.5.2.3	Recall	185
			8.5.2.4	F-score	186
			8.5.2.5	Error rate	186
			8.5.2.6	Time	188
		8.5.3	Discussion		188
	8.6	Conclusion			189

9 Advanced Diabetes Prediction: A Comprehensive Analysis of Machine Learning and Deep Learning Techniques   195
Thottempudi Pardhu, Anwar Bhasha Pattan, Vijay Kumar, Usha Desai, and Biswaranjan Acharya

	9.1	Introduction	196

9.2	Related Work		202
	9.2.1	Literature review on diabetes prediction using machine learning	203
	9.2.2	Literature review on diabetes prediction using deep learning	204
9.3	Discussion		206
	9.3.1	Datasets	206
	9.3.2	Diabetes prediction using machine learning/deep learning techniques	206
9.4	Case Study		213
	9.4.1	Collecting data	213
	9.4.2	Preprocessing	213
	9.4.3	Execution and outcomes	213
	9.4.4	Analysis and examination	214
9.5	Summary		216

10 Intelligent Diagnosis Support System for Screening Diabetes Subjects using Hybrid Machine Learning Algorithms   223

Ch.Rajendra Prasad, Srinivas Samala, Sreedhar Kollem, Ravichander Janapati, Srikanth Yalabaka, and Moola Ramu

10.1	Introduction		224
10.2	Related Work		225
10.3	Materials and Methods		228
	10.3.1	Dataset	228
	10.3.2	Data pre-processing	228
	10.3.3	Machine learning models	230
10.4	Results and Discussion		233
10.5	Conclusion		237

11 Cyber–Physical System for Managing Diabetic Healthcare   241

Usha Desai, Kandala N. V. P. S. Rajesh, T. Kishore Kumar, Varadraj Gurupur, and Ganesh R. Naik

11.1	Background		242
	11.1.1	Epidemiology of DM in India	242
	11.1.2	Epidemiology of DM in USA	243
	11.1.3	Limitations and future directions	245
	11.1.4	Methods to prevent or reduce the effect of DM	245
	11.1.5	Challenges of caregivers of patients suffering from DM	246

11.2	Materials	248
11.3	Methodology	250
11.4	Work Plan	252
	11.4.1 Outcomes of diabetic healthcare system	258
11.5	Conclusion	259

Index   **261**

About the Editors   **263**

Preface

The comprehension of causation is critical in the constantly changing field of diabetes care. This book covers several aspects that are essential to modern healthcare, starting with explaining why it is important to analyze causality for the purpose of managing diabetes. Investigating developments and opportunities made possible by technology, it dives into the cutting edge of digital healthcare systems. Revolutionizing tailored medical procedures, the merging of IoT and expert systems provides personalized continuous diagnostics. Additionally, diabetes treatment has been revolutionized by the use of big data and machine intelligence, which provides insights from the perspectives of both laboratory and clinical personnel. Though early diagnosis is made possible by clever machine learning approaches based on genetic data, specific methodologies such as EfficientNetB3-DTL and optic cup/optic disk segmentation take diagnostics to new heights. Deep learning and machine learning approaches are thoroughly analyzed in order to further develop intelligent methods.

With the number of diabetes cases rising, this book delves deeply into the potential of causal analysis to improve the diabetic care. This book provides a thorough examination of the ways that cutting-edge technologies and data-driven solutions are influencing diabetes care in the future.

The first chapter emphasizes how crucial it is to comprehend the causes of diabetes care so that more focused interventions may be made. The diabetic retinopathy diagnosis systems are carefully dissected, and the complex techniques by which explainable artificial intelligence explains diagnostic paths and provides transparent insights into decision-making processes are explained. The book then delves into the fascinating developments and prospects found in digital diabetes healthcare systems, opening the door for individualized strategies. In-depth analysis of the incorporation of digital platforms into healthcare paradigms is provided in Chapter 2, which also offers a number of prospects for more efficient and individualized patient care.

The Internet of Things (IoT) and expert systems play revolutionary roles in Chapters 3 and 4, and big data and machine intelligence have the potential to completely change the way diabetes is managed and controlled. In order to categorize multiclass diabetic retinal images, Chapter 3 deconstructs the complexity of convolutional neural networks (CNNs), opening the door to precision-driven diagnosis and therapy approaches. In Chapter 4, the emergence of eon technology is the main event. It advocates for a patient-centered strategy that uses technological innovations to enable people to take control of their diabetes, promoting self-management and individualized treatment.

Following this, the book takes a closer look at the impact of these technologies from a laboratory perspective. Intelligent diagnosis support systems assume prominence in Chapter 5, illuminating their pivotal role in augmenting healthcare practitioners' decision-making capabilities through sophisticated screening methodologies. Chapter 6 pioneers the fusion of advanced machine learning with the chronic care model, revolutionizing type-2 diabetes self-management by leveraging futuristic approaches to disease management. Chapter 7 intricately weaves machine intelligence and big data, unveiling their symbiotic relationship and the transformative potential they harbor in offering actionable insights for diabetic care. The disruptive potential of portable healthcare systems is unveiled in Chapter 8, accentuating their ability to alleviate the burdens of diabetes by redefining accessibility and convenience in healthcare delivery.

Finally, the book delves into advanced diabetes prediction techniques and intelligent diagnosis support systems, utilizing both machine learning and deep learning approaches. Chapter 9 meticulously dissects diabetic point-of-care medical systems, shedding light on their real-time monitoring capabilities and their potential in revolutionizing intervention strategies. Intelligent insulin care systems, detailed in Chapter 10, offer a glimpse into precision-driven dosage management, heralding a new era in insulin delivery and patient care. Chapter 11 explains methodology for developing a decision support system that assists diabetes caretakers in monitoring and managing glucose levels, medications, and lifestyle choices, by providing personalized recommendations and actionable insights based on individual patient data management and medication.

This compendium is not merely a repository of technological advancements; it stands as a beacon, guiding researchers, engineers, clinicians, and individuals navigating the intricate pathways of diabetes healthcare towards a future shaped by precision, accessibility, and holistic patient-centric care.

List of Figures

Figure 1.1	Year-wise distribution diabetes patients based on MEPS dataset over the last five years (2017–2021).	3
Figure 1.2	Distribution of diabetes patients age based on MEPS dataset over the last five years (2017–2021).	3
Figure 1.3	Distribution of medical expenses for the top 10 CCSR1X codes based on the MEPS dataset over the last five years (2017–2021).	4
Figure 1.4	Levels and sectors of influence on obesity and diabetes risk (progress in preventing childhood obesity) (National Academies Press, 2007). SES, socioeconomic status.	6
Figure 1.5	CDSS processing pipeline.	7
Figure 2.1	Key elements for implementing digitalization in diabetes.	20
Figure 2.2	Current technologies of digital diabetes care.	22
Figure 2.3	Applications of AI and ML in diabetes management.	23
Figure 2.4	AI/ML-enabled algorithm/medical devices approved by FDA.	24
Figure 2.5	Role of IoT network in diabetes/healthcare.	26
Figure 2.6	Blockchain and health record management.	29
Figure 2.7	Important elements of telemedicine.	32
Figure 2.8	Mobile apps for diabetes care and management.	39
Figure 3.1	Diabetic Type 1, when no or damaged beta cell – so no insulin produced in body.	56
Figure 3.2	Diabetic Type 2, when beta cells are ok – resistance developed against insulin in body organs.	56
Figure 3.3	A general progression of diabetic disease.	58
Figure 3.4	Pre-diabetes-diabetic indication; some of them can be easily registered by IoT sensors.	58
Figure 3.5	A basic structure of IoT.	61
Figure 3.6	A basic structure of the expert system.	62

List of Figures

Figure 3.7	A basic diagram of IoT-expert system based diabetic control management system.	65
Figure 3.8	A basic diagram of developing type-2 and type-1 diabetes, where IoT-based diabetic control management system can start working and an efficient IoT-expert system based model can stop it to proceed further in progressive diabetic stages.	67
Figure 3.9	A modified detailed diagram of progression of the diabetic disease.	71
Figure 4.1	Supervised learning technique for diabetic diagnosis.	85
Figure 4.2	Dynamic processing of decision tree.	88
Figure 4.3	Dynamic processing of decision tree.	89
Figure 4.4	An example methodology of unsupervised learning via clustering.	91
Figure 5.1	The workflow of constructing personalized diabetes treatment regime.	118
Figure 6.1	The architecture of the proposed DR classification through DTL.	130
Figure 6.2	Sample images from the dataset.	131
Figure 6.3	DR dataset samples in each class.	131
Figure 6.4	Pre-processing of H_dr image.	132
Figure 6.5	The proposed EfficientnetB3 DTL structure and its associated parameters.	134
Figure 6.6	Example of IRB.	135
Figure 6.7	Comparison of swish activation function with other activation functions [26].	136
Figure 6.8	Global average pooling.	136
Figure 6.9	Balanced samples for training and testing of the EffientNetB3-DTL model.	138
Figure 6.10	Training and testing accuracies of the proposed EffientNetB3-DTL.	139
Figure 6.11	Training and testing loss of the proposed EffientNetB3-DTL.	140
Figure 6.12	The prediction accuracies of the sample images of each class.	140
Figure 7.1	Retinal fundus.	147
Figure 7.2	Optic disc and optic cup region.	147
Figure 7.3	PPA and optic disc, optic cup region.	150

Figure 7.4	Block diagram of the proposed method.	151
Figure 7.5	Results of active contour technique.	152
Figure 7.6	Results of level set technique.	153
Figure 7.7	Clustering results.	154
Figure 7.8	Segmented optic disc and optic cup.	154
Figure 7.9	CDR results.	156
Figure 8.1	Difficulties in standard prediction method.	174
Figure 8.2	Proposed architecture.	175
Figure 8.3	Processing layers of CBMM.	177
Figure 8.4	The flow diagram of CBMM.	180
Figure 8.5	(a) Training accuracy. (b) Training loss.	183
Figure 8.6	Confusion matrix.	183
Figure 8.7	Precision assessment.	184
Figure 8.8	Accuracy assessment.	185
Figure 8.9	Recall assessment.	186
Figure 8.10	F-score assessment.	187
Figure 8.11	Error rate assessment.	187
Figure 9.1	Usage frequencies of machine learning classifiers.	209
Figure 9.2	Prevalence of deep learning techniques usage.	210
Figure 9.3	Infrequent and underutilized machine learning classifiers in diabetes prediction.	214
Figure 10.1	Proposed model architecture.	229
Figure 10.2	Mean performance metrics with fold cross-validation of the LightGBM.	234
Figure 10.3	ROC curve of the LightGBM.	235
Figure 10.4	Precision−recall curve of the LightGBM.	235
Figure 10.5	Mean performance metrics with fold cross-validation of the LightGBM + KNN.	236
Figure 10.6	ROC curve of the LightGBM.	237
Figure 10.7	Precision−recall curve of the LightGBM.	237
Figure 11.1	Number of articles available in PubMed with the keyword "Wearable glucose monitoring devices for diabetes management."	244
Figure 11.2	Number of systematic review articles on wearable glucose monitoring devices for diabetes management from 2017 to 2023.	244
Figure 11.3	Block diagram of the proposed decision support system for diabetic healthcare.	251

Figure 11.4 Representation of value proposition of proposed work for diabetic health care. 254

Figure 11.5 Technical details of the diabetic healthcare system. 257

List of Tables

Table 3.1	A feedback from individuals regarding diabetic-related health habits.	72
Table 6.1	Hyperparameter setting.	139
Table 6.2	Performance parameters.	140
Table 7.1	Standard baseline data for glaucoma.	156
Table 7.2	Results of cup to disc ratio.	157
Table 7.3	Features selection from GLCM.	159
Table 7.4	Features range.	159
Table 7.5	Results of sensitivity, specificity and accuracy values of different classifiers.	162
Table 8.1	Execution parameters specification.	181
Table 8.2	Database details.	182
Table 8.3	Comparison assessments.	188
Table 8.4	Overall performance of CBMM.	188
Table 9.1	Overview of machine learning algorithm applications.	197
Table 9.2	Comprehensive review of deep learning technique implementations.	201
Table 9.3	Overview of combined model approaches in research.	202
Table 9.4	Overview of datasets utilized in diabetes prediction research.	207
Table 9.5	Comparative analysis of advantages and disadvantages of ML and DL algorithms.	211
Table 9.6	Performance metrics of infrequently used machine learning classifiers in diabetes prediction.	215
Table 10.1	Dataset parameters.	228
Table 10.2	Output classes and features of the dataset.	230
Table 10.3	LightGBM with fivefold cross-validation.	234
Table 10.4	LightGBM and KNN with fivefold cross-validation.	236

List of Contributors

Acharya, Biswaranjan, *Department of Computer Engineering - AI & BDA, Marwadi University, India;*
E-mail: biswaranjan.acharya@marwadieducation.edu.in

Bir, Aritri, *Department of Biochemistry, Dr. B. C. Roy Multi-Speciality Medical Research Centre, IIT Kharagpur, India; E-mail: dr.aritribir@gmail.com*

Chaudhary, Sachin, *School of Computer Science and Applications, IIMT University, India; E-mail: sachin.chaudhary126@gmail.com*

Desai, Usha, *Department of Electronics and Communication Engineering, S.E.A College of Engineering & Technology, Bengaluru, India; E-mail: dr.ushadesai@seaedu.ac.in*

Gerogiannis, Vassilis C., *Department of Digital Systems, University of Thessaly, Greece; E-mail: vgerogian@uth.gr*

Ghosh, Arindam, *Department of Biochemistry, Dr. B. C. Roy Multi-Speciality Medical Research Centre, IIT Kharagpur, India; E-mail: arindam@bcrmrc.iitkgp.ac.in*

Gurupur, Varadraj, *Center for Decision Support Systems and Informatics, School of Global Health Management and Informatics University of Central Florida, USA; E-mail: varadraj.gurupur@ucf.edu*

Jadeja, Nancy, *GCS Medical College, India;*
E-mail: nancyjadeja23@gmail.com

Janapati, Ravichander, *Department of ECE, SR University, India; E-mail: chander3818@gmail.com*

Kanavos, Andreas, *Department of Informatics, Ionian University, Greece;*
E-mail: akanavos@ionio.gr

King, Christian, *Center for Decision Support Systems and Informatics, School of Global Health Management and Informatics University of Central Florida, USA; E-mail: christian.king@ucf.edu*

Kollem, Sreedhar, *Department of ECE, SR University, India; E-mail: ksreedhar829@gmail.com*

Kumar, Bhupendra, *School of Computer Science and Applications, IIMT University, India; E-mail: singhbhupender231@gmail.com*

Kumar, T. Kishore, *Department of Electronics and Communication Engineering, NIT Warangal, India; E-mail: kishoret@nitw.ac.in*

Kumar, Vijay, *SENSE, Vellore Institute of Technology, India; E-mail: vijaykumar@vit.ac.in*

Mayya, Veena, *Center for Decision Support Systems and Informatics, School of Global Health Management and Informatics University of Central Florida, USA; Department of Information & Communication Technology, Manipal Institute of Technology, Manipal Academy of Higher Education (MAHE), India; E-mail: veena.mayya@ucf.edu*

Mishra, Pranjul, *Marwadi University, India; E-mail: pranjulmishra228161@gmail.com*

Naik, Ganesh R., *College of Medicine and Public Health, Flinders University, South Australia; E-mail: ganesh.naik@flinders.edu.au*

Panwar, Rajneesh, *School of Computer Science and Applications, IIMT University, India; E-mail: rajpanwar0710@gmail.com*

Pardhu, Thottempudi, *Department of ECE, BVRIT Hyderabad College of Engineering for Women, India; E-mail: pardhu.t@bvrithyderabad.edu.in*

Patel, Priya, *Department of Pharmaceutical Sciences, Saurashtra University, India; E-mail: patelpriyav@gmail.com*

Pattan, Anwar Bhasha, *Department of ECE, BVRIT Hyderabad College of Engineering for Women, India*

Ponnibala, M., *Department of Biomedical Engineering, Velalar College of Engineering and Technology, India; E-mail: ponnibala@velalarengg.ac.in*

Prasad, Ch.Rajendra, *Department of ECE, SR University, India; E-mail: chrprasad20@gmail.com*

Rajesh, Kandala N. V. P. S., *School of Electronics Engineering, VIT-AP University, India; E-mail: kandala.rajesh2014@gmail.com*

Ramu, Moola, *Department of ECE, Sumathi Reddy Institute of Technology for Women (SRITW), India; E-mail: moola.ramu@gmail.com*

Rao, C. V. Guru, *Department of Computer Science and Engineering Gayatri Vidhya Parishad College of Engineering, Visakhapatnam, India; E-mail: guru_cv_rao@hotmail.com*

Roy, Asitava Deb, *Department of Pathology/Lab Medicine, All India Institute of Medical Sciences, Deoghar, Jharkhand, India; E-mail: asitavadr@gmail.com*

Samala, Srinivas, *Department of ECE, SR University, India; E-mail: srinu486@gmail.com*

Şengöz, Nilgün, *Department of Information Systems and Technologies, Gölhisar School of Applied Sciences, Burdur Mehmet Akif Ersoy University, Turkey; E-mail: nilgunsengoz@mehmetakif.edu.tr*

Sharma, Kewal Krishan, *School of Computer Science and Applications, IIMT University, India; E-mail: drkks57@gmail.com*

Sharmam, Vikas, *School of Computer Science and Applications, IIMT University, India; E-mail: vicky.c610@gmail.com*

Shukla, Madhu, *Marwadi University, India;*
E-mail: madhu.shukla@marwadieducation.edu.in

Srinivas, B., *Department of Computer Science & Engineering (Networks), KITS, India; E-mail: bs.csn@kitsw.ac.in*

Sutariya, Vijaykumar, *Department of Pharmaceutical Sciences, USF Health Taneja College of Pharmacy, University of South Florida, USA; E-mail: vsutariy@usf.edu*

Vadia, Nasir, *Department of Pharmaceutical Sciences, Faculty of Health Sciences, Marwadi University, India; E-mail: nasirvadia@rediff.com*

Vashishth, Tarun Kumar, *School of Computer Science and Applications, IIMT University, India; E-mail: tarunvashishth@gmail.com*

Yalabaka, Srikanth, *Department of ECE, SR University, India; E-mail: srikanthyelabaka7131@gmail.com*

List of Notations and Abbreviations

$\phi(i, j, t)$	Level set function
(i, j)	Coordinates
$\sum_{i=0}^{n}$	Summation operator over i from 1 to M
ξ	Slack variable
C	Regularization parameter
1V1	One-versus-one
1VR	One-versus-rest
ADA	American diabetes association
AHRQ	Agency for healthcare research and quality
AI	Artificial intelligence
ANN	Artificial neural network
AUC	Area under the curve
BHBM	Be healthy, Be mobile
BMI	Body mass index
BNN	Bayesian neural networks
BP	Back-propagation
BR	Bayesian regulation
CART	Classification and regression trees
CBMM	Coati-based multilayer model
CBR	Case-based reasoning
CCSR	Clinical classification software refined
CD	Centralized database
CDC	Centers for disease control and prevention
CDR	Cup to disc ratio
CDSS	Clinical decision support system
CFI	Color fundus imaging
CGM	Continuous glucose monitoring
CNN	Convolutional neural network
CNV	Choroidal neovascularization
COVID	Coronavirus disease
CPS	Cyber–physical system

CVM	Chan-vese model
CV	ChanVese
DCNN	Deep convolutional neural network
DL	Deep learning
DM	Diabetes mellitus
DNA	Deoxyribonucleic acid
DQN	Deep Q-network
DR	Diabetic retinopathy
DRIONS–DB	Digital retinal image for optic nerve segmentation database
DRL	Deep reinforcement learning
DSS	Decision support system
DTL	Deep transfer learning
ECG	Electrocardiograms
eHealth	Electronic health
EHR	Electronic health record
ELM	Extreme learning machine
EX	Exudates
FPR	False positive rate
GAP	Global average pooling
GARIC	Generalized approximate reasoning-based intelligence control
GBDT	Gradient boosting decision tree
GBT	Gradient boosting tree
GDM	Gestational diabetes mellitus
GLCM	Gray level co-occurrence matrix
GRNN	General regression neural network
GrpNN	Grouped neural network
GRU	Gated recurrent unit
H_dr	Healthy diabetic retinopathy
HbA1c	Glycated hemoglobin
HCP	Healthcare professional
HER	Electronic health records
HIPAA	Health insurance portability and accountability act
HM	Hemorrhages
HONN	Higher order neural network
IATV	Interactive television
ICT	Information and communication technology

IDF	International diabetes federation
IDM	Interconnected diabetes management
IDSS	Intelligent diagnosis support systems
IOP	Intraocular pressure
IOP	Increased intraocular pressure
IoT	Internet of things
IOTA	Internet of things application
IRB	Inverted residual blocks
ITU	International telecommunication union
KNN	K-nearest neighbor
LEA	Lower-extremity amputation
LEAs	Lower-extremity amputations
LightGBM	Light gradient boosting machine
LM	Levenberg–marquardt
LMICs	Low- and middle-income countries
LR	Logistic regression
LSTM	Long short-term memory network
M_dr	Moderate diabetic retinopathy
MA	Microaneurysms
MBConv	Mobile inverted bottleneck convolution
mHealth	Mobile health
MHIoT	Medical and healthcare internet of things
Mi_dr	Mild diabetic retinopathy
ML	Machine learning
MLP	Multilayer perceptron
mNGS	Metagenomic next-generation sequencing
MVP	Minimum viable product
NASA	National aeronautics and space administration
NB	Naive bayes
NDPP	National diabetes prevention programme
NIDDK	National institute of diabetes and digestive and kidney diseases
NLP	Natural language processing
NN	Neural network
OCT	Optical coherence tomography
OD	Optic disk
ONH	Optic nerve head
P_dr	Proliferative diabetic retinopathy
PCA	Principal component analysis

PDA	Patient decision aid
PDAs	Patient decision aids
PID	Pima indian dataset
PIDD	Pima indian diabetes database
PNN	Probabilistic neural network
PPA	Peri papillary atrophy
RBF	Radial basis function
ReLU	Rectified linear unit
REP	Reduced error pruning
RF	Random forest
RFID	Radio frequency identification
RGB	Red green blue
RIM-ONE	Retinal image database for optic nerve evaluation
RMSE	Root mean squared error
RNN	Recurrent neural network
ROC	Receiver operating characteristic
ROI	Region of interest
S_dr	Severe diabetic retinopathy
SCG	Scaled conjugate gradient
SES	Socioeconomic status
SHAP	Shapley additive explanations
SLFN	Single-hidden layer feedforward neural networks
SMO	Sequential minimal optimization
SMS	Short message service
SOM	Self-organizing map
STARPAHC	Space technology applied to rural papago advanced health care
SVC	Soft voting classifier
SVG	Support vector regression
SVM	Support vector machine
t	Time
T1DM	Type-1 diabetes mellitus
T2DM	Type-2 diabetes mellitus
TPR	True positive rate
UCI	University of California Irvine
US FDA US	Food and drug administration
VCD	Vertical cup diameter

VDD	Vertical disc diameter
VP	Value proposition
WHO	World health organisation
XAI	Explainable artificial intelligence
XGBoost	eXtreme gradient boosting

1
Importance of Analyzing Causality for Diabetes Care

Veena Mayya[1,2], Christian King[1], and Varadraj Gurupur[1]

[1]Center for Decision Support Systems and Informatics,
School of Global Health Management and Informatics,
University of Central Florida, USA
[2]Department of Information & Communication Technology,
Manipal Institute of Technology,
Manipal Academy of Higher Education (MAHE), India
E-mail: veena.mayya@ucf.edu; christian.king@ucf.edu;
varadraj.gurupur@ucf.edu

Abstract

In the realm of health, causality often embodies multifaceted and complex relationships, encapsulating a blend of genetic, environmental, and lifestyle determinants. Delving into these causal pathways and elucidating the causes and consequences within diabetes care bears paramount significance for a multitude of reasons. The identification of causal attributes associated with diabetes has the potential to aid in the prevention of this metabolic disorder. Moreover, understanding causality can engender personalized care, tailoring treatment strategies to meet the unique needs of individuals afflicted with diabetes. Beyond individual care, the insight derived from causality can bolster the efficacy of treatment outcomes and play an instrumental role in the innovation of new pharmaceuticals and therapeutic modalities. On a societal scale, comprehending the cause-and-effect relationships in diabetes can inform policy development, thereby enhancing public health initiatives. Through a thorough understanding of the causal dynamics underlying diabetes, educational and awareness programs can be designed with increased efficacy, thereby enabling individuals to make more informed decisions pertaining to their health. Consequently, a detailed, multifaceted exploration of these causal relationships is vital for efficacious diabetes management and prevention.

Keywords: Diabetes mellitus, causality analysis, clinical decision, support systems, health informatics.

1.1 Prevalence of Diabetes

Diabetes mellitus (or diabetes), a metabolic disorder resulting in hyperglycemia, affects an increasing proportion of the global population. According to a report by the Centers for Disease Control and Prevention (CDC) [1], approximately 37.3 million Americans (representing 11.3% of the population) were diagnosed with diabetes, with an additional 8.5 million individuals likely undiagnosed. This chronic disease is associated with considerable mortality, having been implicated in over 100,000 deaths in 2021 alone. Particularly concerning is the high prevalence among the elderly, with the American Diabetes Association (ADA) indicating that in 2023, over a quarter (26%) of Americans aged 65 and older were diagnosed with this condition [2].

To analyze the significance of the prevalence of diabetes in the United States, data from the Medical Expenditure Panel Survey spanning the last five years (2017–2021) was collated. The datasets from each of these five years were consolidated and merged with corresponding medical condition files using the person ID as a key identifier. Specifically, the diabetes diagnosis (DIABDX) key was employed to track the number of individuals diagnosed with diabetes over the years. As depicted in Figure 1.1, there is a noticeable increase in the number of patients diagnosed with diabetes over this time period. Furthermore, the person's age last time eligible (AGELAST) variable was utilized to plot the age-wise distribution of diabetes patients as depicted in Figure 1.2.

Diabetes is categorized into several forms, including type-1 diabetes, type-2 diabetes, and gestational diabetes, each exhibiting unique etiologies and prevalence rates. Gestational diabetes is a transitory condition affecting a minority (2%–10%) of pregnancies annually. Type-1 diabetes, an autoimmune condition resulting in the destruction of pancreatic insulin-producing cells, can manifest at any age but is usually diagnosed in children and young adults. However, it is relatively rare, accounting for only 5%–10% of all diagnosed cases. In stark contrast, type-2 diabetes is the most common form,

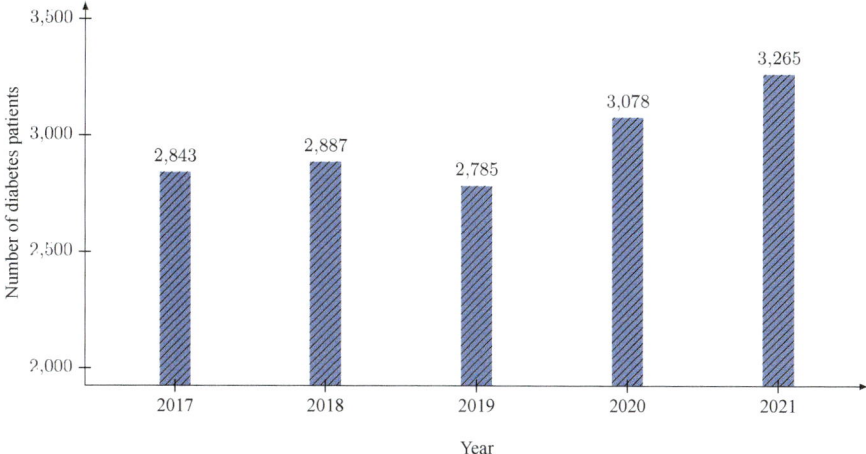

Figure 1.1 Year-wise distribution diabetes patients based on MEPS dataset over the last five years (2017–2021).

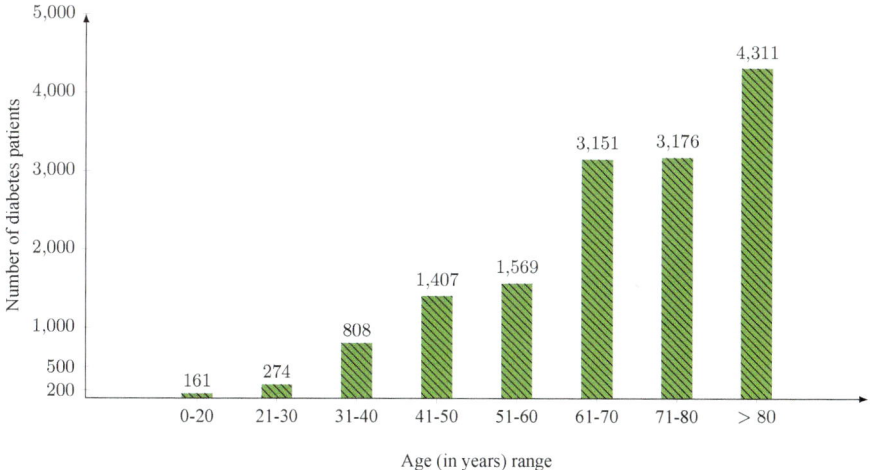

Figure 1.2 Distribution of diabetes patients age based on MEPS dataset over the last five years (2017–2021).

comprising 90%–95% of all cases. It affects primarily adults and an increasing number of children, characterized by insulin resistance or insufficient insulin production.

The continuing rise in diabetes, particularly type-2, represents a significant public health challenge in the United States. This concern is accentuated

4 *Importance of Analyzing Causality for Diabetes Care*

by the serious complications associated with poor disease management, including cardiovascular disease [3–8], stroke [9–11], kidney disease [12–15], ocular disease [16–19], and lower limb amputations [20–22]. Specifically, diabetic retinopathy has become the leading cause of adult blindness, with the National Eye Institute predicting a twofold increase from 7.7 million affected Americans in 2010 to 14.6 million by 2050 [23]. Moreover, foot complications from diabetes are globally recognized as major causes of patient disability. Lower-extremity amputations (LEAs), a prevalent outcome, significantly impact patients' functional abilities and overall life quality.

Based on the collated MEPS 2017–2021 data, diabetes is identified as the fourth most significant source of financial burden in terms of medical expenses in the United States. Figure 1.3 provides a visual representation of the comprehensive healthcare expenditures attributed to various diseases, employing the Clinical Classification Software Refined (CCSR) from the collated MEPS dataset. CCCSR consolidates International Classification of Diseases, 10th Revision, Clinical Modification/Procedure Coding System

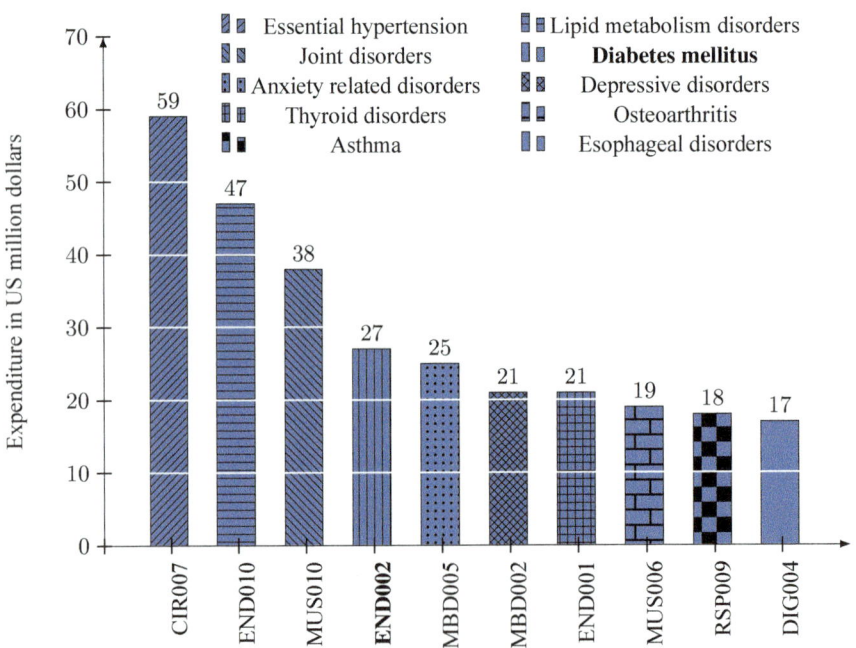

Figure 1.3 Distribution of medical expenses for the top 10 CCSR1X codes based on the MEPS dataset over the last five years (2017–2021).

(ICD-10-CM/PCS) codes into clinically meaningful categories. Figure 1.3 reveals a notable allocation of approximately 27 million US dollars specifically dedicated to addressing the healthcare needs associated with diabetes. Expenditures related to other complications stemming from diabetes are not encompassed within these expenditures. This substantial financial outlay underscores the significant economic impact imposed by the disease on the healthcare system.

Consequently, there arises a compelling imperative to direct strategic efforts toward curbing the escalating prevalence of diabetes and, concurrently, alleviating the financial strain linked to managing its repercussions. Recognizing the pivotal role of preventive measures and interventions in tackling this public health challenge, prioritizing initiatives aimed at both preventing the onset of diabetes and optimizing management strategies becomes imperative. By proactively addressing the root causes and implementing effective management protocols, healthcare systems can endeavor to mitigate the burgeoning financial toll inflicted by diabetes, thereby fostering improved health outcomes and resource utilization within the broader public health landscape.

1.2 Factors Contributing to Diabetes

There are multiple factors contributing to diabetes and elevated levels of glucose. Earlier literature has focused on individual risk factors such as education, income, and job occupation [24]. Subsequent literature [25] has recognized the need to examine the disease through a social and environmental perspective. Different ecological models have been proposed to examine the determinants of diabetes [26]. One such representation is the socioecological model proposed by the Committee on Progress in Preventing Childhood Obesity (National Academies Press, 2007). This model offers a comprehensive understanding of how diverse environments may interact and influence the risk of diabetes [27] (refer Figure 1.4).

A large body of literature has shown that neighborhoods and the neighborhood context have a large influence on health outcomes, including diabetes [28]. The composition of neighborhoods by race, ethnicity, and socioeconomic status affects neighborhood environments, leading to health disparities and inequities. Additionally, the physical attributes of neighborhoods strongly affect residents' health. For instance, dilapidated physical conditions can discourage physical activity [29]. Likewise, neighborhood violence or safety concerns can adversely affect physical and mental health

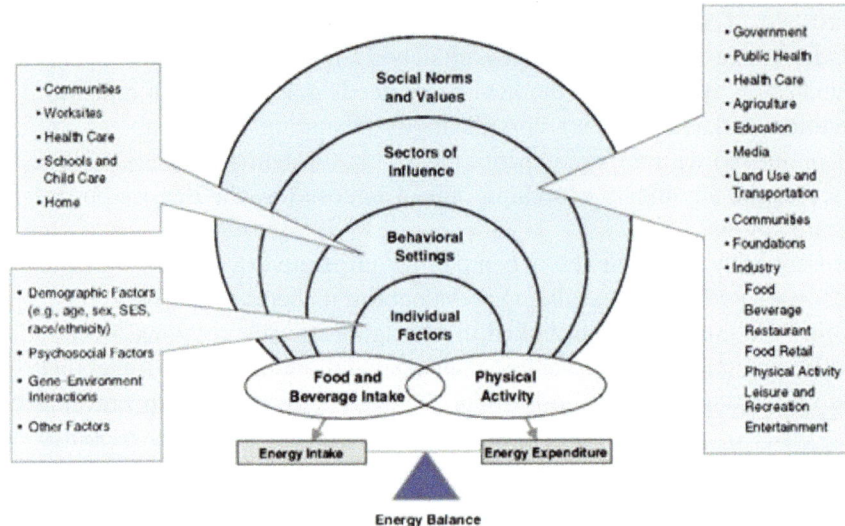

Figure 1.4 Levels and sectors of influence on obesity and diabetes risk (progress in preventing childhood obesity) (National Academies Press, 2007). SES, socioeconomic status.

[30, 31]. Furthermore, these conditions can lead to decreased physical activity and increased stress, potentially contributing to depression. The food environment, characterized by the availability, accessibility, and distribution of food sources, also plays a critical role [1]. A number of studies link the prevalence of fast-food outlets and convenience stores in a neighborhood to a higher incidence of diabetes [32, 33]. The consumption of energy-dense, nutritionally poor food prevalent in these establishments can lead to increased diabetes rates. Lastly, disadvantaged neighborhoods often house individuals of lower socioeconomic status who are more likely to experience food insecurity [34, 35]. Food insecurity, defined as insufficient access to food for a healthy, active life, may lead to malnutrition and/or undernutrition. This, in turn, increases the risk of developing diabetes [36].

1.3 Decision Support Tools and Diabetes Management

Decision support tools have been used in different ways to improve the treatment of diabetes. The first one is predicting whether a patient has diabetes. This is important because 23% of all adults with diabetes have an undiagnosed condition and are unaware that they have this chronic condition. As a result, early detection of the condition could lead to better treatment and quality of life for patients. Increasingly, clinical decision support systems (CDSS)

1.3 Decision Support Tools and Diabetes Management

are utilized in the early detection of diabetes. These systems incorporate information from electronic health records, laboratory tests, medical images, and other sources in order to predict the risk of developing diabetes [37–39]. The machine learning algorithms implemented into these systems can learn from immense quantities of data to recognize patterns that human clinicians might overlook, thereby enhancing their early detection capabilities [40–42]. Recently deep-learning-based decision support systems are able to predict diabetes illness from patient data with high accuracy [38, 42–44]. Patient decision aids (PDAs) can also play a role in early detection by encouraging healthy lifestyle changes and monitoring symptoms that may indicate the onset of diabetes [45]. These aids can assist at-risk individuals in making well-informed decisions regarding their health behaviors, potentially delaying or preventing the onset of diabetes. The development of wearable devices and mobile applications that monitor parameters such as glucose levels, physical activity, and dietary intake is attributable to the rise of digital health technologies. These can provide useful information for the early detection of diabetes and promote self-management of health [46–48]. Telemedicine platforms, which provide remote consultations and health examinations, can also aid in the early detection of diabetes. Regular monitoring through these platforms can help identify alterations in health status that signal the onset of diabetes, allowing for timely intervention [49–53].

The majority of computer-aided CDSS follow a common processing pipeline, as illustrated in Figure 1.5. The data collection phase involves gathering and storing relevant medical data in accordance with the specified task and labeling the data based on expert clinician opinions. This phase is typically the most time-consuming, especially in prospective studies that require data collection during the ongoing study period. In contrast, retrospective studies, which analyze existing patient cohorts, often have shorter timelines for data collection. Regardless of the study type, the collected data must be labeled by expert clinicians for training the CDSS. Due to differences in observation, both inter- and intra-observer variability can introduce noise into the labeled data. Additionally, the raw data may contain missing values,

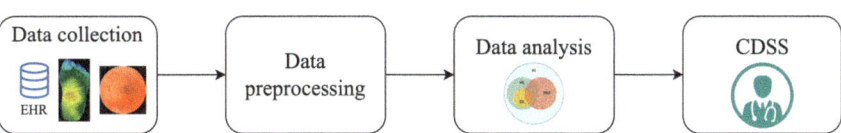

Figure 1.5 CDSS processing pipeline.

a large number of noisy features, unwanted image regions, or an imbalance in the dataset. Before training the CDSS, the data must be preprocessed to address these issues. Once the data is cleaned, it is analyzed using various statistical, supervised, unsupervised, ontology-based, and knowledge-based algorithms relevant to the clinical task. This analysis also involves optimizing parameters and hyperparameters for the specific task. If the analysis yields feasible results with validation data, the CDSS can be deployed for further fine-tuning and to assist in clinical decision-making.

Another way in which decision support tools are used in diabetes research is through the management of the disease. For example, one study used clinical decision support to optimize hospital glycemic management to effectively reduce hyperglycemic events among hospitalized patients [54]. Pérez-Gandía et al. [55] used a decision support system to help type-1 diabetic patients ensure that they follow a proper management regimen for their condition. Once patients check their glucose prediction, they can use the information to decide whether to correct their level of hyperglycemia or their level of hypoglycemia. Another study used decision support systems to develop an algorithm that provides weekly insulin dosage recommendations to adults with type-1 diabetes using multiple daily injections. The decision support system algorithm allowed for early identification of dangerous insulin regimens and may be used to improve glycemic outcomes among people with type-1 diabetes [56].

In summary, decision support system tools have the potential to improve the detection of diabetes and the management of the condition, thus improving the health outcomes of patients with diabetes. By integrating diverse sources of health data and employing advanced analysis techniques such as machine and deep learning, these tools can identify patterns that predict the onset of diabetes more efficiently than traditional methods. In addition, enhanced diabetes management facilitated by these decision support tools can result in reduced hospitalizations, a crucial factor in the current healthcare environment where the emphasis is on reducing hospital readmission rates. Importantly, these developments can also contribute to a decrease in diabetes-related mortality rates. Thus, the collective effect of decision support system tools extends beyond individual patient care, influencing broader healthcare outcomes and influencing the future of diabetes management. Thus, the collective effect of decision support system tools extends beyond individual patient care, influencing broader healthcare outcomes and influencing the future of diabetes management.

1.4 Conclusions

Understanding and analyzing causality are of the utmost importance in diabetes management. The chronic nature of diabetes necessitates a thorough understanding of the causal relationships between genetics, lifestyle, and the environment. This information not only aids in the early detection and diagnosis of diabetes, but also in the development of personalized treatment plans tailored to the specific requirements and circumstances of each individual patient. Modern technologies such as machine learning and decision support systems have substantially enhanced our ability to analyze these causal relationships. Using the power of large datasets, these technologies enable the identification of subtle patterns and trends that would otherwise go unnoticed. Moreover, knowledge of causality has significant implications for preventative measures. By identifying and addressing the underlying causes of diabetes, we can implement interventions that promote healthier communities at the population level. Thus, the analysis of causality is not merely a theoretical exercise, but also a practical tool with the potential to radically enhance diabetes care outcomes.

With technological advancements, our ability to understand and analyze causal relationships will further improve. There is potential for causality analysis to improve diabetes management. By comprehending how various factors interact to influence blood glucose levels, we can create more effective management strategies. For instance, we may be able to anticipate how changes in diet or exercise will impact the blood glucose levels of an individual, allowing for more precise and individualized management of the condition. The future of analyzing causality for diabetes care is promising, with the potential to dramatically improve our ability to prevent, detect, and manage diabetes.

References

[1] (2023) Centers for disease control and prevention, national diabetes statistics report. https://www.cdc.gov/diabetes/data/statistics/statistics-report.html, [Online; accessed 10-July-2023]

[2] (2023) Statistics about diabetes, american diabetes association. https://www.diabetes.org/resources/statistics/statistics-about-diabetes, [Online; accessed 10-July-2023]

[3] Caussy C, Aubin A, Loomba R (2021) The relationship between type 2 diabetes, nafld, and cardiovascular risk. Current Diabetes Reports 21(5).

[4] Eckel RH, Bornfeldt KE, Goldberg IJ (2021) Cardiovascular disease in diabetes, beyond glucose. Cell Metabolism 33(8):1519–1545.
[5] Shah A, Isath A, Aronow WS (2022) Cardiovascular complications of diabetes. Expert Review of Endocrinology and Metabolism 17(5):383–388.
[6] Miller RG, Costacou T (2022) Cardiovascular disease in adults with type 1 diabetes: Looking beyond glycemic control. Current Cardiology Reports 24(10):1467–1475.
[7] Wong ND, Sattar N (2023) Cardiovascular risk in diabetes mellitus: epidemiology, assessment and prevention. Nature Reviews Cardiology.
[8] Motairek I, Al-Kindi S (2023) Ameliorating cardiovascular risk in patients with type 2 diabetes. Endocrinology and Metabolism Clinics of North America 52(1):135–147.
[9] Kissela B, Air E (2006) Diabetes: Impact on stroke risk and poststroke recovery. Seminars in Neurology 26(1):100–107.
[10] Bell DS, Goncalves E (2020) Stroke in the patient with diabetes (part 1) âĂŞ epidemiology, etiology, therapy and prognosis. Diabetes Research and Clinical Practice 164.
[11] Krinock MJ, Singhal NS (2021) Diabetes, stroke, and neuroresilience: looking beyond hyperglycemia. Annals of the New York Academy of Sciences 1495(1):78–98.
[12] Thomas S (2010) Diabetic nephropathy. Medicine 38(12):639–643.
[13] Piccoli GB, Clari R, Ghiotto S, Castelluccia N, Colombi N, Mauro G, Tavassoli E, Melluzza C, Cabiddu G, Gernone G, Mongilardi E, Ferraresi M, Rolfo A, Todros T (2013) Type 1 diabetes, diabetic nephropathy, and pregnancy: A systematic review and meta-study. Review of Diabetic Studies 10(1):6–26.
[14] Krolewski AS, Skupien J, Rossing P, Warram JH (2017) Fast renal decline to end-stage renal disease: an unrecognized feature of nephropathy in diabetes. Kidney International 91(6):1300–1311.
[15] Li J, Li X, Gathirua-Mwangi W, Song Y (2020) Prevalence and trends in dietary supplement use among us adults with diabetes: The national health and nutrition examination surveys, 1999-2014. BMJ Open Diabetes Research and Care 8(1),
[16] Purushothaman I, Zagon IS, Sassani JW, McLaughlin PJ (2021) Ocular surface complications in diabetes: The interrelationship between insulin and enkephalin. Biochemical Pharmacology 192.
[17] Liu F, Liu C, Lee IXY, Lin MTY, Liu YC (2023) Corneal dendritic cells in diabetes mellitus: A narrative review. Frontiers in Endocrinology 14

[18] Kropp M, Golubnitschaja O, Mazurakova A, Koklesova L, Sargheini N, Vo TTKS, de Clerck E, Polivka J, Potuznik P, Polivka J, Stetkarova I, Kubatka P, Thumann G (2023) Diabetic retinopathy as the leading cause of blindness and early predictor of cascading complicationsâĂŤrisks and mitigation. EPMA Journal 14(1):21–42.

[19] Priyadarsini S, Whelchel A, Nicholas S, Sharif R, Riaz K, Karamichos D (2020) Diabetic keratopathy: Insights and challenges. Survey of Ophthalmology 65(5):513–529.

[20] Walicka M, RaczyÅDska M, Marcinkowska K, Lisicka I, Czaicki A, Wierzba W, Franek E (2021) Amputations of lower limb in subjects with diabetes mellitus: Reasons and 30-day mortality. Journal of Diabetes Research 2021.

[21] Tork HM, Tajbakhsh F (2015) A survey of the presence of lower limb arterial insufficiency in diabetic patients attending the surgery department of golestan hospital for reasons other than vascular diseases. Biomedical and Pharmacology Journal 8(2):1201–1208.

[22] Navarro-Flores E, Cauli O (2020) Quality of life in individuals with diabetic foot syndrome. Endocrine, Metabolic and Immune Disorders - Drug Targets 20(9):1365–1372.

[23] (2023) Facts about diabetic eye disease, national eye institute. https://www.nei.nih.gov/learn-about-eye-health/eye-conditions-and-diseases/diabetic-retinopathy, [Online; accessed 10-July-2023]

[24] Hill-Briggs F, Adler NE, Berkowitz SA, Chin MH, Gary-Webb TL, Navas-Acien A, Thornton PL, Haire-Joshu D (2021) Social determinants of health and diabetes: a scientific review. Diabetes care 44(1):258.

[25] Walker RJ, Gebregziabher M, Martin-Harris B, Egede LE (2014) Relationship between social determinants of health and processes and outcomes in adults with type 2 diabetes: validation of a conceptual framework. BMC endocrine disorders 14:1–10.

[26] Hill-Briggs F, Adler NE, Berkowitz SA, Chin MH, Gary-Webb TL, Navas-Acien A, Thornton PL, Haire-Joshu D (2021) Social determinants of health and diabetes: A scientific review. Diabetes Care 44(1):258–279.

[27] Hill JO, Galloway JM, Goley A, Marrero DG, Minners R, Montgomery B, Peterson GE, Ratner RE, Sanchez E, Aroda VR (2013) Scientific statement: Socioecological determinants of prediabetes and type 2 diabetes. Diabetes Care 36(8):2430–2439.

[28] Diez Roux AV, Mair C (2010) Neighborhoods and health. Annals of the New York Academy of Sciences 1186:125–145.

[29] Rees-Punia E, Hathaway ED, Gay JL (2018) Crime, perceived safety, and physical activity: A meta-analysis. Preventive medicine 111:307–313.
[30] Clark C, Ryan L, Kawachi I, Canner MJ, Berkman L, Wright RJ (2008) Witnessing community violence in residential neighborhoods: A mental health hazard for urban women. Journal of Urban Health 85(1):22–38.
[31] Franzese RJ, Covey HC, Tucker AS, McCoy L, Menard S (2014) Adolescent exposure to violence and adult physical and mental health problems. Child Abuse & Neglect 38(12):1955–1965,
[32] Kanchi R, Lopez P, Rummo PE, Lee DC, Adhikari S, Schwartz MD, Avramovic S, Siegel KR, Rolka DB, Imperatore G, et al. (2021) Longitudinal analysis of neighborhood food environment and diabetes risk in the veterans administration diabetes risk cohort. JAMA network open 4(10):e2130789–e2130789
[33] Mazidi M, Speakman JR (2018) Association of fast-food and full-service restaurant densities with mortality from cardiovascular disease and stroke, and the prevalence of diabetes mellitus. Journal of the American Heart Association 7(11):e007651
[34] Denney JT, Kimbro RT, Sharp G (2018) Neighborhoods and food insecurity in households with young children: A disadvantage paradox? Social Problems 65(3):342–359.
[35] Morrissey TW, Oellerich D, Meade E, Simms J, Stock A (2016) Neighborhood poverty and children's food insecurity. Children and Youth Services Review 66:85–93.
[36] Gucciardi E, Vahabi M, Norris N, Del Monte JP, Farnum C (2014) The intersection between food insecurity and diabetes: a review. Current nutrition reports 3:324–332.
[37] Holmes A, Todd A (2012) Evidence-based clinical decision support systems for the prediction and detection of three disease states in critical care. The Knowledge Engineering Review 27(4):487–500.
[38] Rabie O, Alghazzawi D, Asghar J, Saddozai FK, Asghar MZ (2022) A decision support system for diagnosing diabetes using deep neural network. Frontiers in Public Health 10:861062.
[39] Yahyaoui A, Jamil A, Rasheed J, Yesiltepe M (2019) A decision support system for diabetes prediction using machine learning and deep learning techniques. In: 2019 1st International informatics and software engineering conference (UBMYK), IEEE, pp 1–4.
[40] Rajkomar A, Dean J, Kohane I (2019) Machine learning in medicine. New England Journal of Medicine 380(14):1347–1358.

[41] Chaki J, Thillai Ganesh S, Cidham S, Ananda Theertan S (2022) Machine learning and artificial intelligence based diabetes mellitus detection and self-management: A systematic review. Journal of King Saud University - Computer and Information Sciences 34(6):3204–3225.

[42] Sharma T, Shah M (2021) A comprehensive review of machine learning techniques on diabetes detection. Visual Computing for Industry, Biomedicine, and Art 4(1),

[43] Zhao D, Wang W, Tang T, Zhang YY, Yu C (2023) Current progress in artificial intelligence-assisted medical image analysis for chronic kidney disease: A literature review. Computational and Structural Biotechnology Journal 21:3315–3326.

[44] Triantafyllidis A, Kondylakis H, Katehakis D, Kouroubali A, Koumakis L, Marias K, Alexiadis A, Votis K, Tzovaras D (2022) Deep learning in mhealth for cardiovascular disease, diabetes, and cancer: Systematic review. JMIR mHealth and uHealth 10(4).

[45] Wilson E, Kenny A, Dickson-Swift V (2021) Feasibility of implementing patient decision aids in a rural victorian health service. Rural Remote Health 21(2):6187.

[46] Wang Q, Li H, Wang X, Wu D, Chen J (2018) Using a self-monitoring application to improve medication adherence in patients with kidney disease: a randomized controlled trial. Patient Preference and Adherence 12:2023–2030.

[47] Chmayssem A, Nadolska M, Tubbs E, Sadowska K, Vadgma P, Shitanda I, Tsujimura S, Lattach Y, Peacock M, Tingry S, Marinesco S, Mailley P, Lablanche S, Benhamou PY, Zebda A (2023) Insight into continuous glucose monitoring: from medical basics to commercialized devices. Microchimica Acta 190(5).

[48] Gopalan HS, Haque I, Ahmad S, Gaur A, Misra A (2019) "diabetes care at doorsteps": A customised mobile van for the prevention, screening, detection and management of diabetes in the urban underprivileged populations of delhi. Diabetes and Metabolic Syndrome: Clinical Research and Reviews 13(6):3105–3112.

[49] Posadzki P, Mastellos N, Ryan R, Gunn LH, Felix LM, Pappas Y, Gagnon MP, Julious SA, Xiang L, Oldenburg B, Car J (2016) Automated telephone communication systems for preventive healthcare and management of long-term conditions. Cochrane Database of Systematic Reviews 2016(12).

[50] Demaerschalk BM, Vegunta S, Vargas BB, Wu Q, Channer DD, Hentz JG (2017) Reliability of real-time video smartphone for assessing national institutes of health stroke scale scores in acute stroke patients. Stroke 48(12):3305–3309.
[51] Kruklitis R, Miller M, Valeriano L, Shine S, Opstbaum N, Chestnut V (2022) Applications of remote patient monitoring. Primary Care - Clinics in Office Practice 49(4):543–555.
[52] Ben-Assuli O (2022) Measuring the cost-effectiveness of using telehealth for diabetes management: A narrative review of methods and findings. International Journal of Medical Informatics 163.
[53] Bhatt P, Liu J, Gong Y, Wang J, Guo Y (2022) Emerging artificial intelligence-empowered mhealth: Scoping review. JMIR mHealth and uHealth 10(6).
[54] Pichardo-Lowden A, Umpierrez G, Lehman EB, Bolton MD, DeFlitch CJ, Chinchilli VM, Haidet PM (2021) Clinical decision support to improve management of diabetes and dysglycemia in the hospital: a path to optimizing practice and outcomes. BMJ Open Diabetes Research and Care 9(1):e001557.
[55] Pérez-Gandía C, García-Sáez G, Subías D, Rodríguez-Herrero A, Gómez EJ, Rigla M, Hernando ME (2018) Decision support in diabetes care: the challenge of supporting patients in their daily living using a mobile glucose predictor. Journal of diabetes science and technology 12(2):243–250.
[56] Tyler NS, Mosquera-Lopez CM, Wilson LM, Dodier RH, Branigan DL, Gabo VB, Guillot FH, Hilts WW, El Youssef J, Castle JR, et al. (2020) An artificial intelligence decision support system for the management of type 1 diabetes. Nature metabolism 2(7):612–619.

2

Advances and Opportunities in Digital Diabetic Healthcare Systems

Nasir Vadia[1], Priya Patel[2], and Vijaykumar Sutariya[3]

[1]Department of Pharmaceutical Sciences, Faculty of Health Sciences, Marwadi University, India
[2]Department of Pharmaceutical Sciences, Saurashtra University, India
[3]Department of Pharmaceutical Sciences, USF Health Taneja College of Pharmacy, University of South Florida, USA
E-mail: nasirvadia@rediff.com; patelpriyav@gmail.com; vsutariy@usf.edu

Abstract

Diabetes has far-reaching implications, not only for those directly affected but also for caregivers, families, and communities. Unhealthy eating and sedentary lifestyles fuel the diabetes surge. Challenges to effectively controlling blood sugar in patients stem from limited access to specialized care, poor medication adherence, and inadequate self-care motivation. Nonetheless, healthcare is rapidly transforming, with state-of-the-art digital technologies like smartphone apps, telemedicine, m-health, device integration, machine learning, and AI poised to reshape diabetes management. These innovations offer great potential to improve care efficacy and encourage patients to play a more active self-care role. These innovations hold immense potential to enhance the efficacy of diabetes management while empowering patients to take a more active role in their self-care. The current healthcare system has given some great instances of overhauling the diabetes ecosystem by utilizing the possibilities of digital transformation. A revolutionary shift in diabetes management via digital transformation offers tremendous potential for optimizing healthcare cost and resource efficiency. This approach promises to elevate the quality and uniformity of diabetes care, transcending geographical constraints and also contributing to enhanced governance and policy formulation across diverse healthcare sectors. Furthermore, it catalyzes encouraging innovation and fostering collaboration among industry stakeholders, ultimately leading to the development of innovative products and solutions that support diabetes management and care. To create a better, more attainable future for individuals with diabetes, this chapter focuses on the most recent digital and technological developments and breakthroughs to identify the appropriate device for diabetes monitoring and management.

Keywords: Digital health, medical data, mobile healthcare, monitoring sensors, computer decision support, diabetes therapy.

2.1 Background

Global diabetes rates have surged, with India ranking second after China [1]. The Indian Council of Medical Research's India Diabetes report reveals a current diabetic population of 77 million. Moreover, India faces significant instances of prediabetes and undiagnosed diabetes [2, 3]. Despite advancements in diabetes treatments and technologies, a substantial proportion of individuals with diabetes encounter challenges in achieving improved glycemic outcomes [4, 5]. Because of the disorder's complicated expression, persons who have diabetes need customized medical support. However, the limitations of conventional approaches' limited individual care and healthcare resources have made it difficult to implement a personalized strategy to manage diabetes. Diabetes management presents a host of hurdles for patients, necessitating strategic approaches to ensure optimal glycemic, blood pressure, and lipid control while minimizing complications. The foremost challenges encompass five critical aspects: (1) optimizing current therapies to maximize efficacy, (2) empowering patients through education to self-manage their condition, (3) enhancing treatment adherence, (4) streamlining timely diagnosis and early insulin initiation, and (5) redefining healthcare provision for those with chronic conditions. Addressing these challenges promises to reshape the landscape of diabetes care and improve patient outcomes significantly [6–8]. In the daily lives of modern people, information and communication technology (ICT) development has resulted in significant changes. Over the years, a remarkable transformation has unfolded, originating from the advent of personal computers and further fuelled by the wired internet. Nevertheless, it is the rapid advancement of smartphones and wireless technologies that has ushered in even more substantial changes.

In the contemporary environment, fundamental elements of ICT have not merely become essential facets of daily existence but have also demonstrated substantial utility in the realms of medical and healthcare provisions [9, 10]. Moreover, the importance of evaluating and regulating daily routines, including nutrition, exercise, and medication adherence, cannot be overstated. Embracing various digital health components within the chronic disease arena is projected to be widespread, and as a result, holds the potential to profoundly revolutionize the current medical paradigm [11].

The National Digital Health Mission marks a pivotal stride in India's Digital India program, signifying the government's dedicated efforts to embrace digitalization in the healthcare sector and tackle complex health challenges, such as diabetes [12]. India, the world's second-largest mobile phone market utilizes smartphone proliferation to bolster e-governance efficacy. Digital technology integration is anticipated to drive profound shifts in healthcare, spanning health promotion, prevention, primary care, and specialized treatments. This heralds an era of comprehensive and attainable healthcare in India, despite resource constraints stemming from its vast population. Given the persistent nature of diabetes, continuous monitoring is imperative for effective glycemic control and pre-empting future complications. Here, digital technology offers the patient, the doctor, and the policymaker the best tracking and management option. Due to its ability to facilitate consumer involvement, behavior modification, and impact analytics, digital technology has proven helpful in preventing and treating diabetes [6]. These cutting-edge technologies offer seamless data acquisition and comprehension, presenting a promising future in diabetes care. Among the latest innovations are NovoPen 6 and NovoPen Echo Plus, pioneering connected insulin pens currently in development, poised to hit the market this year. These intelligent pens establish a connection with smartphones to automate the gathering of insulin dosage information, effectively closing the information loop between glucose and insulin level measurements. Through integration with a linked blood glucose meter or continuous glucose monitor, these pens offer the potential to simplify the management of diabetes. Particularly, the sphere of type-2 diabetes care can experience substantial advantages from these innovative digital approaches, underscoring their exceptional promise. Embracing this transformative era of digital healthcare is crucial in improving the lives of millions living with diabetes. The utilization of digital diabetes management solutions has witnessed a substantial rise and is projected to maintain its upward trajectory. As per 2020 studies, the number of individuals using diabetes apps globally is anticipated to reach 23.5 million in 2022, a significant surge from 6.1 million in 2019. However, relying solely on technology falls short of meeting the comprehensive needs of diabetes patients. To unlock the full potential of digitization for both healthcare systems and individuals with diabetes, fostering stronger collaboration among payers, providers, and healthcare professionals (HCPs) is essential. A united effort toward an integrated strategy for diabetes care is imperative to ensure that those with diabetes receive the level of support they truly require.

A holistic approach is advocated, encompassing medicine, nutrition, and mental health, to provide a transformative experience for diabetes management. The integration of meters and apps in diabetes care poses challenges for patients and healthcare professionals, necessitating effective control measures to minimize burdens. Interconnected diabetes management (IDM) emerges as a promising approach, utilizing digital solutions to streamline various facets of diabetes treatment. IDM aims to improve patient care by enhancing communication among healthcare systems and those involved in the treatment process. Health2Sync, an example of an IDM program, gathers and shares vital diabetes-related information, offering personalized tips and reminders. Research suggests that all-encompassing digital management strategies can enhance type-2 diabetes outcomes [13]. Likewise, in a separate investigation, four key findings emerged from a 2020 survey executed by the U.S. Department of Health and Human Services Agency for Healthcare Research and Quality (AHRQ). This survey focused on assessing the efficiency, user-friendliness, and functionalities of smartphone applications designed for diabetes management and available in the commercial market [14]. Furthermore, studies on smartphone applications for diabetes management reveal potential benefits in A1C reduction, though more evidence is required to ascertain their long-term efficacy. Overall, embracing IDM and addressing user concerns may facilitate better diabetes care through digital solutions [15, 16]. Roughly 50% of individuals who initially acquire health-focused applications discontinue their usage due to factors such as the time investment needed for data input, waning enthusiasm, undisclosed costs, usability challenges, and concerns regarding the sharing of personal data [18, 19].

2.2 Transformation of Diabetes Management from Conventional to Digital

The realm of digital health is undergoing a transformative wave powered by internet connectivity, social media, and mobile technology. With billions of internet and mobile users globally, these innovative networks offer unprecedented global outreach, especially in remote regions facing healthcare challenges. Embracing digital health solutions can bridge these gaps and revolutionize healthcare accessibility and quality on a large scale. These tools are particularly beneficial for individuals living with diabetes, enhancing training, patient engagement, and information exchange for better prevention and treatment. Cutting-edge technology enables close monitoring of physical activity, dietary patterns, and blood glucose levels, fostering collaboration between healthcare practitioners and patients. The potential of mobile technology in mitigating non-communicable diseases, especially for diabetes patients in low- and middle-income countries (LMICs), is exemplified by initiatives like the World Health Organisation (WHO) and International Telecommunication Union's (ITU) "Be Healthy, Be Mobile" (BHBM). Utilizing basic SMS interventions has already proven effective in reducing diabetes-related comorbidities. Integration of digital tools into comprehensive diabetes strategies is currently underway to further enhance care quality within LMICs [20]. The dynamic evolution of digital diabetes education and care presents revolutionary prospects for dietitians, propelled by ongoing technological advancements. The emergence of integrated delivery systems, patient-centered medical homes, and accountable care organizations serves as prime examples of this transformative change. These shifts empower dietitians to spearhead the redesign of care processes and adopt value-based reimbursement models. Their expertise in diabetes treatment and patient-generated health data utilization positions diabetes-focused dietitians and educators to assume mentorship and leadership roles. This explores the potential impact of digital diabetes care on the dietitian's role and its implications for reshaping healthcare. Dietitians can act as distribution conduits for digital treatments, establishing comprehensive feedback loops to improve diabetes self-management. Effective readiness and proficiency in leveraging patient-generated health data are crucial, necessitating proper training and expertise in population-level data utilization for quality improvement and data-driven clinical interventions [21, 22]. Innovative digital diabetes tools have empowered patients and improved disease self-management, bridging the gap between doctor visits [23]. This technology, including continuous glucose monitoring, insulin pumps, and smartphone apps, allows individuals to efficiently regulate their glucose levels, prevent complications, and enhance their quality of life [24, 25]. Advancements like the Internet of Things (IoT), AI, and machine learning are promising for diabetic management [26]. The rise of eHealth and mHealth, utilizing portable electronic

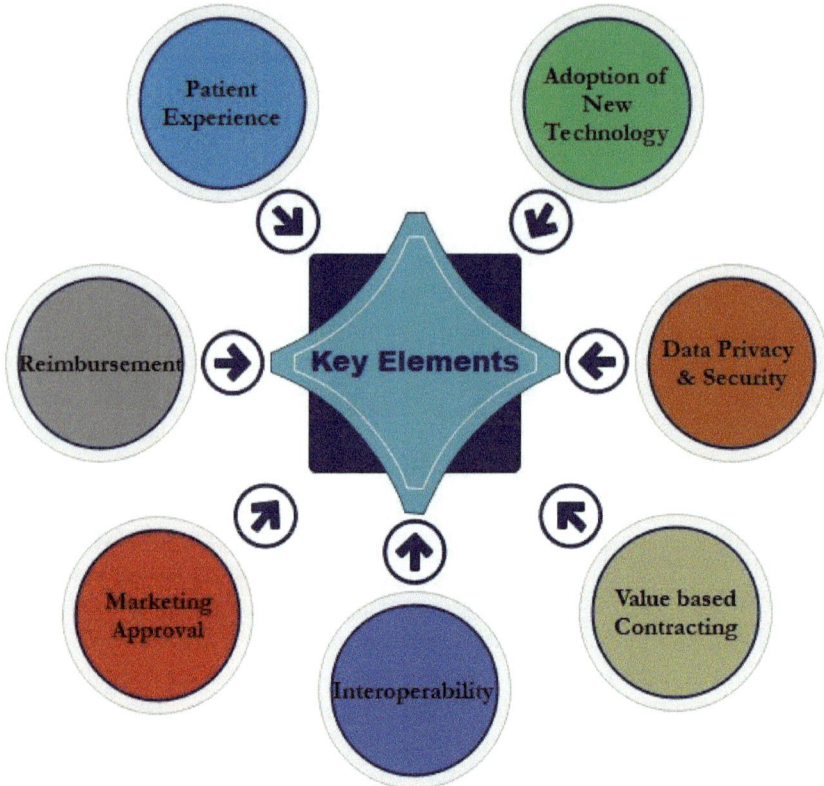

Figure 2.1 Key elements for implementing digitalization in diabetes.

devices, has revolutionized medical information delivery and healthcare services [27]. Since 2013, digital tools have been transforming treatment strategies, enabling personalized care and better communication for medical professionals and patients [28, 29]. Experts predict a surge in adopting digital tools for diabetes care in the next five years, including telemedicine, education platforms, treatment apps, and glucose management solutions [30, 31]. Some key elements need to be assessed before implementing digital diabetes (Figure 2.1).

2.3 Digital Technologies for Diabetes Management

In light of the increasing diabetes prevalence and a global shortage of healthcare professionals, innovative approaches to diabetes care delivery are

2.3 Digital Technologies for Diabetes Management

urgently needed. This transformation aims to expand access to care, alleviate patient burdens, improve healthcare system efficiency, and ease financial strain on health systems and payers. A novel digital diabetes concept, leveraging cutting-edge technologies, holds promise in addressing these challenges effectively. Current obstacles to achieving better clinical outcomes for diabetic patients include therapeutic inertia and lack of accessible, standardized, clinically relevant data. Digital diabetes technologies that collect, transfer, and analyze crucial diabetic data autonomously can break down these barriers. Patient empowerment and self-care are essential in diabetes management, and various digital tools and apps, incorporating motivational tools, incentives from social media, and gamification strategies, have shown potential to revolutionize diabetes care and improve treatment outcomes significantly. The term "digital health" encompasses a broad spectrum of technologies and platforms that advance life sciences, clinical practices, healthier lifestyles, and the management of chronic diseases like diabetes. Key players include social media, smartphones, mobile apps, wearables, cloud-based data systems, and evidence-based research [32, 33]. This diverse technological landscape has revolutionized digital health, driven by the rise of social media, smartphones, apps, wearables, cloud data, and research evidence [34, 35]. A description of a few notable digital technologies utilized in diabetes care today (Figure 2.2) will follow to further highlight the technology's key features.

2.3.1 Artificial intelligence and machine learning (AI and ML)

Artificial intelligence (AI) and machine learning (ML) have initiated a transformative era in computer systems, simulating human intelligence and cognitive abilities. Their rapid evolution and broad applications across industries have made healthcare the most promising domain[36, 37]. This study delves into the transformative influence of AI and ML in the medical field, emphasizing applications such as disorder detection, health tracking, patient data administration, drug innovation, surgery, telemedicine, medical statistics, tailored therapy, and imaging. Notably, AI integration in primary care bolsters clinical judgments, diagnostics, practice administration, and practitioner education. Pritesh Mistry's classification underscores AI's capacity for proactive identification of undiagnosed ailments, predictive modeling for health results, and advanced clinical decision-making [38].

Cutting-edge applications in diabetes care include automated retinal screening, predictive risk stratification, and patient self-management tools,

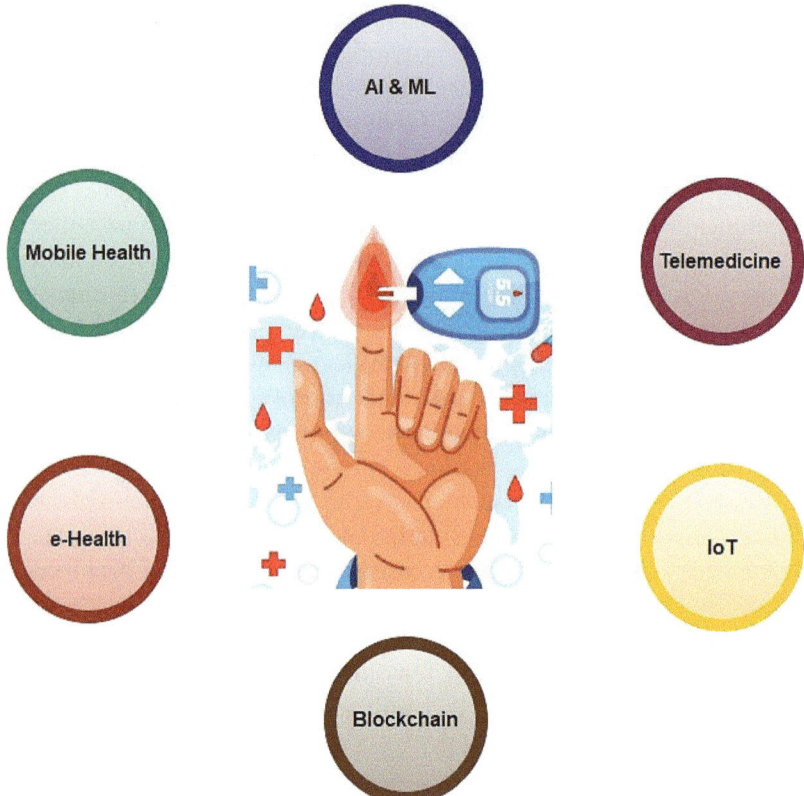

Figure 2.2 Current technologies of digital diabetes care.

with early detection of diabetic retinopathy standing out as a cost-effective measure to prevent avoidable blindness. Leveraging image-based screening for diabetic foot ulcers and retinal abnormalities improves intervention and enhances the quality of life for diabetes patients [39]. The immense potential of AI and ML in healthcare promises more efficient and personalized patient care and advances the field of medical sciences. The substantial advances in AI approaches, particularly in case-based reasoning (CBR), have revolutionised the management of diabetes. The integration of advanced machine learning algorithms, including a k-nearest neighbor, support vector machine, artificial neural network, naive Bayes, decision tree, random forest, and regression trees, has resulted in an automated system for screening blood glucose fluctuations. This system utilizes outlier removal techniques, cross-validation methods, and a range of classifiers such as

2.3 Digital Technologies for Diabetes Management

linear discriminant analysis, quadratic discriminant analysis, Gaussian process classification, and logistic regression. By incorporating feature selection strategies, the system effectively identifies individuals susceptible to diabetes based on their genetic and metabolic characteristics [40, 41]. Additionally, support vector regression (SVG) has emerged as a promising method for diabetes treatment, facilitating the creation of a novel hypoglycemia predictor, which generates early alerts for preventive interventions during critically low blood glucose levels. This groundbreaking digital solution in diabetes care has the potential to transform management practices and improve patient outcomes. The remarkable scope of AI and ML applications in the field is evidenced by the approval and availability of over 520 medical algorithms driven by AI by the US Food and Drug Administration (US FDA) in the US market as of January 2023. Some important application areas of AI and ML in diabetes care are illustrated in Figure 2.3.

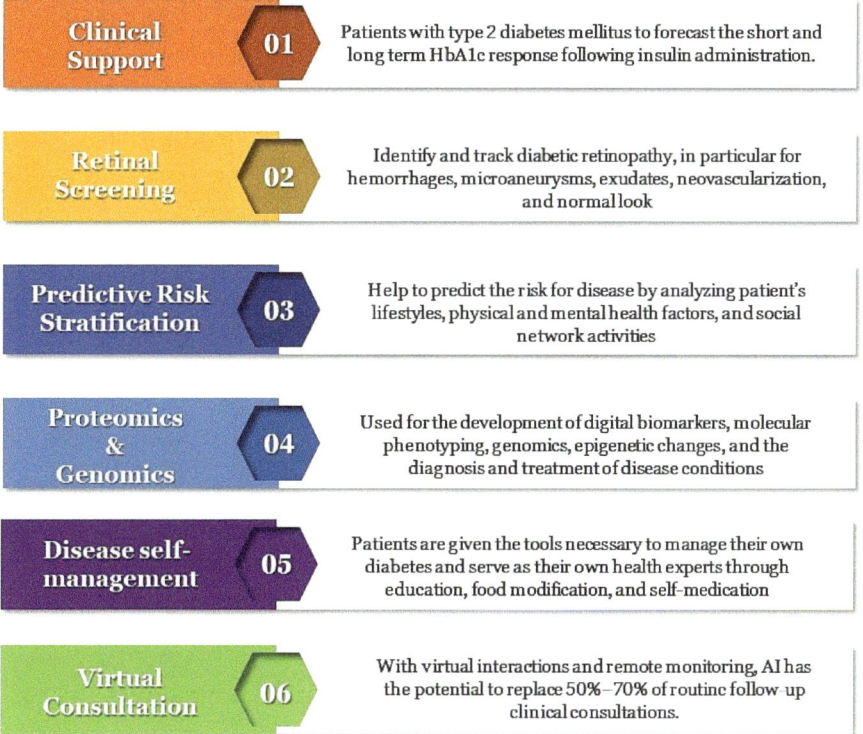

Figure 2.3 Applications of AI and ML in diabetes management.

AI- and ML-based FDA-approved medical devices/algorithms are presented in Figure 2.4. AI and ML can be used in diabetes care to improve patient outcomes and optimize treatment plans.

Here are some ways that AI and ML can be applied in diabetes care:

- Personalized treatment plans: Cutting-edge AI and ML technologies have demonstrated their potential in efficiently analyzing vast troves of patient data. These groundbreaking approaches enable the generation of customized treatment strategies tailored to each unique patient, thereby enhancing the efficacy of treatments while concurrently mitigating the likelihood of complications.
- Early detection of complications: These technologies offer the potential to thoroughly scrutinize patient data, enabling the timely identification of early indications of diabetes-related ailments like neuropathy and retinopathy. By empowering healthcare professionals to take prompt action, this innovative approach aids in curtailing the advancement of these complications and improving patient outcomes significantly.
- Improved patient monitoring: AI and ML have paved the way for real-time monitoring of patient data, facilitating timely alerts to healthcare providers concerning fluctuations in patient health. By leveraging these technologies, potential medical emergencies can be averted, and early intervention becomes possible, ultimately enhancing patient care and safety.

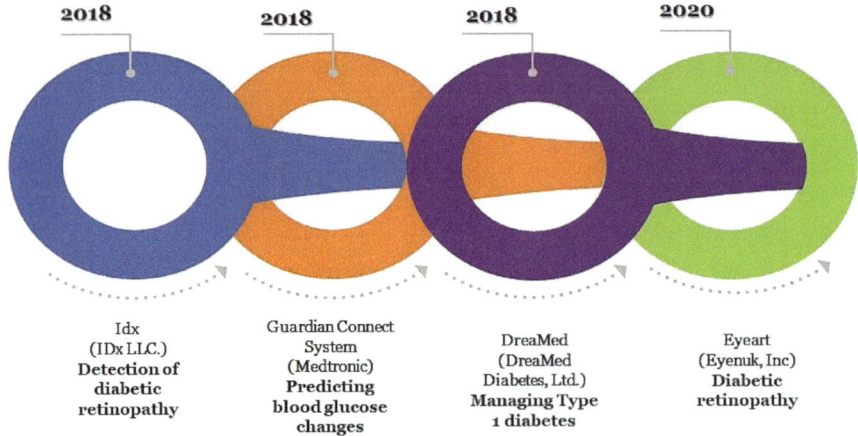

Figure 2.4 AI/ML-enabled algorithm/medical devices approved by FDA.

- Predictive modeling: These techniques enable the creation of robust predictive models, offering healthcare providers the ability to accurately detect individuals at a heightened risk of developing diabetes or its related complications. This revolutionary advancement empowers proactive measures to avert diabetes onset and timely intervention strategies to safeguard patients from potential complications.
- Drug discovery: AI and ML can be used to analyze large datasets to identify potential new drug targets for diabetes treatment. This can help to accelerate the drug discovery process and improve the efficacy of diabetes treatments.

In general, the utilization of AI and ML holds immense promise in elevating the standard of diabetes care. By facilitating personalized treatment strategies, detecting complications at an early stage, enhancing patient monitoring, employing predictive modeling, and expediting drug discovery processes, these advanced technologies offer tremendous opportunities for advancements in diabetic healthcare.

2.3.2 Medical and healthcare Internet of Things (MHIoT)

The Internet of Things (IoT) represents a cutting-edge advancement in communication, seamlessly bridging the digital world with everyday sensors and functional devices. Embracing an all-IP-based architecture, IoT empowers the connection between physical and virtual entities, revolutionizing communication across diverse domains [42]. In the realm of healthcare, IoT adoption has led to the development of intelligent healthcare ecosystems that integrate advanced sensing technologies with complementary hardware and software components, enabling real-time monitoring and management of patients [43, 44]. Leveraging real-time data from IoT sensors, wearable devices, and smartphones, IoT-driven medical devices offer superior support for managing diabetes, enabling continuous monitoring of crucial parameters such as calorie intake, physical activity, insulin sensitivity, and blood circulation. Harnessing the power of big data and artificial intelligence further enhances the potential of IoT-based medical devices, democratizing comfortable healthcare and prioritizing patient well-being. By combining wireless technology and machine learning, clinicians can thoroughly analyze patient records and make informed recommendations, utilizing various IoMT devices in therapeutic fields [45–47]. The study showcases a computer-based method that imitates the pancreas' actions through a mathematical model, while revolutionary advancements in IoT are expected to be game-changers. Specifically, a "continuous glucose monitoring" device is introduced, administering appropriate insulin based on an optimum glycemia equilibrium model. Leveraging IoT devices and AI, the research demonstrates impressive accuracy in diabetes treatment, enabling real-time assessments and widespread dissemination of results. Sensor-derived data, along with radio frequency identification (RFID) technology, promises to elevate healthcare delivery in clinics and homes, enhancing disease surveillance and predictive modeling for effective containment [48]. The revolutionary promise of the IoT lies in its potential to transform diabetes care through continuous health monitoring, empowering healthcare providers with real-time patient data. Figure 2.5 illustrates the role of IoT networks in diabetes management.

Here are some specific applications of IoT in diabetes care:

- Continuous glucose monitoring: IoT devices offer continuous patient blood glucose monitoring, supplying real-time data to healthcare providers, and facilitating timely interventions.
- Wearable devices: IoT-driven wearables, such as intelligent wristwatches and health trackers, present an opportunity to continuously track patients' physical activity, sleep behaviors, and various health indicators. This valuable information holds the potential to craft individualized treatment strategies and enhance overall patient well-being.

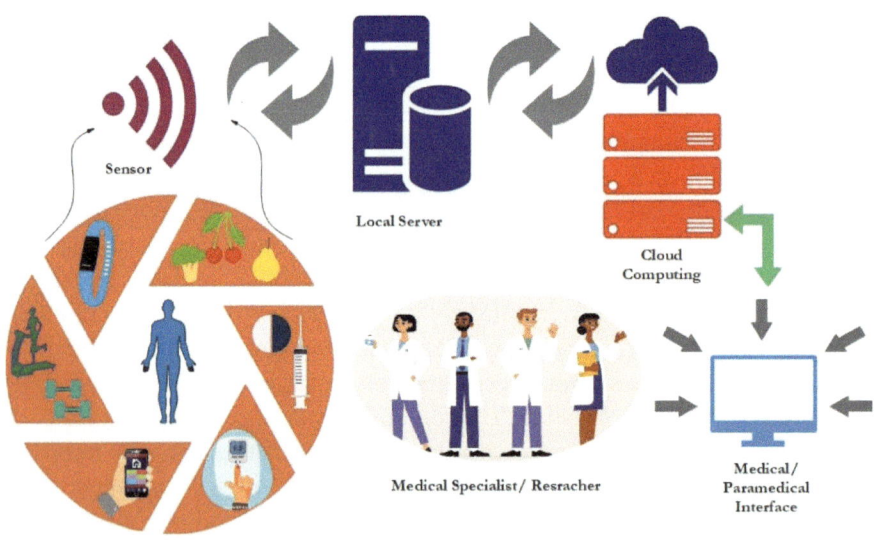

Figure 2.5 Role of IoT network in diabetes/healthcare.

- Remote patient monitoring: Utilizing IoT devices allows for remote monitoring of patients' well-being, empowering healthcare providers with real-time access to track and promptly intervene when needed. This transformative approach has the potential to significantly curtail hospital readmissions and elevate overall patient care and outcomes.
- Smart insulin pens: IoT-enabled insulin pens can be used to track insulin dosing and deliver timely medication reminders to patients to enhance medication adherence, thereby mitigating insulin dosing errors and associated complications.
- Telemedicine: IoT devices can be used to enable telemedicine consultations between patients and healthcare providers, enhancing care accessibility while minimizing the necessity for in-person consultations.

In essence, the IoT holds immense promise in revolutionizing diabetes care, as it facilitates the seamless monitoring of patients' health, offering instantaneous data to healthcare practitioners, and empowering timely interventions when required. This transformative capability has the potential to significantly enhance patient well-being, leading to improved outcomes, and a marked reduction in the incidence of diabetes-related complications.

2.3.3 Blockchain

Blockchain, a decentralized and open-source digital ledger, revolutionizes industries with its secure and unchangeable transaction recording across multiple computers. Each validated block is linked to the previous one, creating an indisputable chain of data. Transparency is guaranteed as all transactions are openly recorded and verified, ensuring high accountability. By distributing data across networks instead of centralized databases, Blockchain improves stability and minimizes vulnerability to cyberattacks. In healthcare, this technology offers manufacturers a powerful tool to monitor product utilization, enhancing transparency and reliability [49]. Notably, it combats counterfeit medications by ensuring complete traceability [50], fortifying the integrity of the medical supply chain. Furthermore, blockchain acts as an impregnable repository for medical histories, safeguarding patient information [51]. With tamper-proof records, patient confidentiality is ensured, solidifying the potential of blockchain to transform healthcare with unparalleled data security [52–54].

The healthcare industry generates vast amounts of information daily in diverse formats, encompassing records, economic documents, clinical test results, imaging tests, and vital sign evaluations. The expansion of healthcare databases has been challenged by issues concerning data access and collection. However, innovative solutions are emerging, with blockchain technology emerging as a promising approach. Blockchain holds the potential to enhance data authenticity, verification, and dissemination within healthcare networks, leading to profound improvements in expenditures, data quality, and healthcare delivery. This transformative technology offers a transparent, decentralized network without intermediaries, revolutionizing the future of healthcare information management. This presents a revolutionary approach to health data management through the integration of cutting-edge blockchain transactions within data lakes. These state-of-the-art blockchain-based repositories securely store vast arrays of medical information, incorporating a user's unique identification code, encrypted health records, and precise timestamps for transaction origination. The data lakes offer unparalleled scalability, accommodating diverse data formats such as images, documents, and key-value stores. Authorized healthcare professionals gain exclusive access to the data lake, while patients retain control over their data through blockchain-managed decryption keys.

This robust and streamlined system facilitates a wide range of health research investigations, personalized treatment enhancements guided by genetic indicators, and the determination of factors impacting preventive healthcare. The integration of blockchain technology ensures the legitimacy of medical records, with each file or image digitally signed by responsible medical professionals before encryption and storage. Patients receive notifications about additions to their blockchain, while a mobile dashboard application empowers them to monitor access, control data sharing, and flexibly manage access permissions. This groundbreaking approach empowers users with precise and comprehensive control over their health data, marking a significant leap forward in healthcare data management. The integration of blockchain in patient data management is illustrated in Figure 2.6. Blockchain tech presents unparalleled potential to revolutionize healthcare, notably in diabetes care and medical data handling. This explores the potential of leading blockchain platforms, including Ethereum, NEO, NEM, and EOSIO, in revolutionizing diabetes care through medical data management, drug traceability, and remote patient monitoring via IoT applications. Notably, NEM stands out as a robust platform, enabling easy interaction through simple web calls and accommodating multiple clinical facilities seeking to contribute and retrieve patient data. The platform's incorporation of multi-signature contracts ensures enhanced data security by requiring consent from linked parties for account access and modifications. Furthermore, blockchain-based

2.3 Digital Technologies for Diabetes Management

platforms like HealthMudra and MedChain utilize optimization and machine learning algorithms to provide diabetes prevention recommendations and empower patients to control their medical records securely [55, 56]. Hyperledger Fabric and IOTA demonstrate promise in monitoring patients' vital signs and securely exchanging physiological data from wearables and IoT devices [57, 58]. Overall, blockchain technology establishes a secure, decentralized data storage and sharing platform, fostering patient privacy, seamless collaboration, and ultimately enhancing patient outcomes in diabetes care [59]. Some potential advantages of using blockchain in diabetes care include:

- Secure and decentralized data storage: One of the biggest advantages of blockchain is its secure and decentralized data storage system. Blockchain technology enables secure patient data storage, obviating the need for a central governing body, reducing the threat of data breaches, and maintaining continuous authorized access.
- Improved patient privacy: Blockchain can help improve patient privacy by giving patients control over their data. Patients have the authority to select healthcare providers and researchers with data access privileges, which can be revoked at their discretion. This empowers patients to

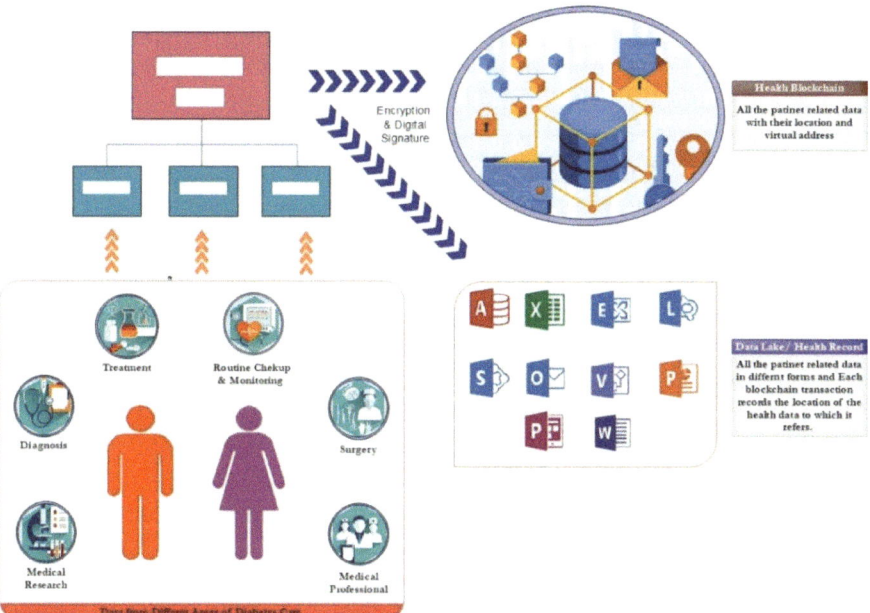

Figure 2.6 Blockchain and health record management.

thwart unauthorized data access, thereby fortifying patient data security and confidentiality.
- Enhanced data sharing and collaboration: Blockchain enables seamless data sharing and collaboration among healthcare providers and researchers. This secure and decentralized platform fosters interconnectivity, dismantling silos and promoting the exchange of patient data across diverse healthcare organizations.
- Improved clinical trials: Utilizing blockchain technology can enhance clinical trials through its transparent and unchangeable ledger of trial data, thus safeguarding against data tampering and ensuring utmost accuracy and reliability.
- Enhanced patient outcomes: Blockchain's transformative potential lies in fostering data sharing and collaboration, thereby elevating patient outcomes in diabetes care. Through seamless access to comprehensive patient data, healthcare providers gain vital insights, enabling informed treatment decisions and personalized high-quality care.

2.3.4 Telemedicine

Diabetes progression can lead to various complications, necessitating a collaborative approach for effective management. A well-rounded diabetes management team, comprising healthcare professionals from diverse disciplines, plays a pivotal role in providing comprehensive care. However, geographical dispersion can hinder the team's effectiveness. Regular reviews following medical consultations can improve patient outcomes significantly. Telemedicine, combined with decision support systems, enables efficient healthcare delivery, enhancing provider–patient communication and optimizing schedules. Moreover, telemedicine promotes electronic health records and health management information systems, facilitating seamless routine healthcare data collection [60]. Such digital health initiatives can address several health system challenges and lead to improved diabetes management on a larger scale. Telemedicine encompasses any medical activity involving a distance component, and the terms "telehealth" and "telemedicine" have been occasionally used interchangeably [61]. The success of telemedicine relies heavily on the efficient electronic transmission of medical data among remote locations. Despite its recent surge in interest, telemedicine has a history spanning over six decades, with early development significantly influenced by NASA's efforts in the 1960s during space missions. Notable milestones include the STARPAHC initiative, providing medical assistance

2.3 Digital Technologies for Diabetes Management 31

to the Papago Indian Reservation and astronauts in orbit, and the North-West Telemedicine Project in Australia. In 1989, the pioneering Space Bridge to Armenia/Ufa program enabled telemedicine consultations between Armenia and the US using video, voice, and facsimile technologies [62, 63]. These initiatives mark significant milestones in global telemedicine advancements. This technology covers the seamless integration of hardware, software, and communication channels to facilitate efficient information exchange and remote consultations between geographically distant sites. Vital hardware components, including computers, printers, scanners, and videoconferencing systems, are pivotal in facilitating the capture and transmission of patient data encompassing images, reports, and videos. This article delves into two key telemedicine technologies: the "store and forward" approach, suited for non-emergency scenarios, and the "face-to-face" consultation utilizing interactive television (IATV) innovation [64]. A myriad of medical specialties, spanning teledermatology, telepathology, teleradiology, psychiatry, internal medicine, rehabilitation, cardiology, pediatrics, obstetrics and gynecology, and neurology, reap rewards from these telemedicine applications. Furthermore, inventive systems like teleneuropsychology, telenursing, telepharmacy, and telerehabilitation are examined, underscoring their substantial contributions to patient care and remote healthcare services [65]. Teleneuropsychology allows remote neuropsychological consultations and assessments, while telenursing facilitates nursing care from a distance.

Telepharmacy provides medical advice and medication monitoring when direct interaction with a pharmacist is not possible. Telerehabilitation employs advanced communication technologies for clinical assessments and therapy, improving outcomes for rehabilitation patients. Overall, this publication sheds light on the diverse applications of telemedicine, underscoring its potential to revolutionize modern healthcare. Primitive elements of telemedicine are presented in Figure 2.7. Some critical telemedicine-based diabetes management technologies are described below.

2.3.4.1 Telephonic/mobile care

New studies validate that enhancing patient involvement through regular phone communication enhances motivation, medication adherence, and metabolic control. Capitalizing on the ubiquity and usage of mobile phones, these interventions can be swiftly applied. In addition to distributing a therapy management plan diary during in-person visits, clinicians can offer valuable recommendations for blood glucose self-monitoring, insulin therapy, and dietary or physical activity guidelines. We can assess the effectiveness of

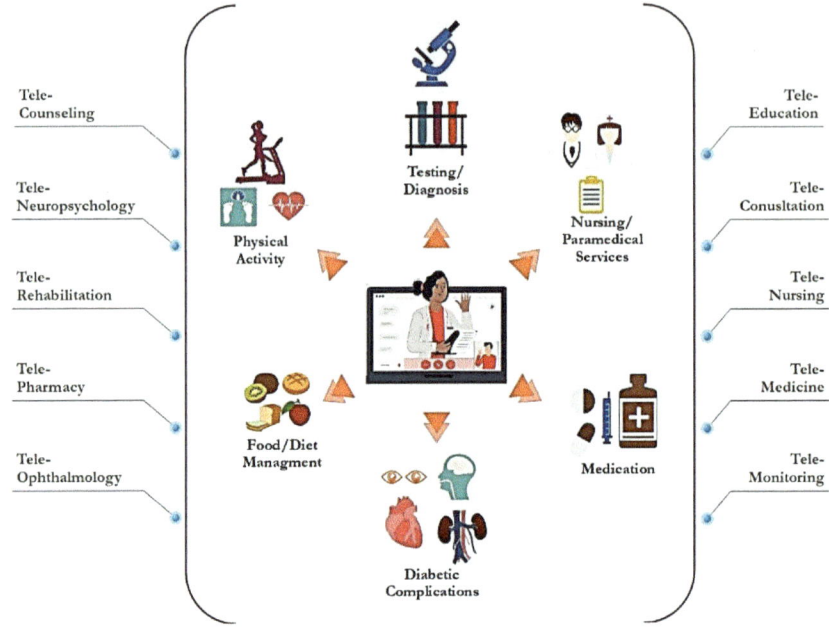

Figure 2.7 Important elements of telemedicine.

these measures implementation by making follow-up phone calls. In addition, a hotline number may be provided, enabling the patient to call ahead of time in case of an emergency, device issues, or other diabetes-related questions. Video conversations may also be appropriate for in-person consultations with specific patients who want tangible proof of the healthcare provider's identity to develop confidence and trust in those supervising their daily therapy. In a similar vein, a video call allows the medical expert to gauge the patient's motivation and attitude more accurately [66]. Video calls provide a valuable advantage by enabling accurate assessment of glucose data, traditionally recorded in paper diaries. Contemporary clinical practice has seen a rise in virtual training sessions delivered via phone or video conferences. These sessions provide remote instruction on diabetes topics including device usage, dietary recommendations, and behavior adjustments. The landscape of diabetes management apps has greatly grown, exceeding 1500 offerings across app stores. These apps, spanning diverse platforms, providers, and devices, now constitute the largest category of health-related applications [67–69]. The innovative tele diabetic care system integrates basal/bolus insulin documentation, carbohydrate tracking guidance, and automated feedback on glucose trends. This cutting-edge approach enables efficient assessment of potential diabetes-related issues, facilitates personalized therapy, and ensures effective glucose management. Video calls play a pivotal role in visually evaluating conditions like diabetic foot syndrome, skin complications resulting from insulin administration or continuous glucose monitoring, as well as technical concerns related to insulin delivery devices or glucose monitors. Embracing this advanced tele diabetic care system promises to revolutionize diabetes management with far-reaching impacts on patient outcomes. Web-based consultations for diabetic foot ulcers' remote monitoring exhibit remarkable efficacy in enhancing wound healing and averting complications. Utilizing advanced temperature-sensing techniques like infrared thermography, liquid-crystal thermography, and thermistor-based sensors at different foot points enables early detection of potential ulcers, prompting timely clinical examination and reducing unnecessary patient visits. Moreover, telemedical-based monitoring of diabetic foot issues is expected to benefit significantly from innovative approaches like tissue oxygenation with hyperspectral imaging and photographic techniques.

2.3.4.2 Systems for continuous glucose monitoring

Proficient insulin therapy users with diabetes are now greatly advantaged by the widespread adoption of continuous glucose monitoring (CGM) systems. These systems utilize glucose sensors to offer real-time interstitial fluid readings, facilitated by their increasing global acceptance, enhanced precision, and positive reimbursement scenarios across nations [70]. By transferring glucose values to a reader or smartphone, or through scanned data extraction, the need for frequent capillary glucose readings is significantly reduced. Consequently, treatment adherence, glycemic control, and overall quality of life have seen remarkable improvements with the adoption of CGM devices. Seamless incorporation of insulin administration and glucose level monitoring is achievable, either via compatible devices within a single brand or through cross-device communication facilitated by software applications such as Diasend. Such tools empower both diabetes patients and healthcare professionals to easily visualize and discuss continuous glucose monitoring data, insulin dosages, and dietary intake. Notably, this approach has proven effective for elderly and pediatric patients as well [71].

2.3.4.3 Diabetic retinopathy control techniques

After diabetes diagnosis, vision impairments often remain asymptomatic, underscoring the importance of frequent screening for diabetic retinopathy. Telemedicine offers a well-suited solution for ophthalmology due to its visual and image-intensive nature, making telescreening in early diabetic retinopathy stages highly advisable. Various basic retinal cameras have been developed, enabling qualified non-ophthalmologist personnel to perform telescreening. Particularly in regions with limited resources or access to conventional screening programs, telemedicine-based approaches have been successfully implemented, facilitating the identification of individuals in need of ophthalmological care [72].

2.3.4.4 Insulin pen

An efficient technique for providing care to children, patients with cognitive impairment, and those with poor therapy adherence involves the use of wireless gadgets. These devices send data on insulin dosing frequency and timing to mobile applications, seamlessly documenting the information. Smart pens enhance users' understanding of dosing habits, aiding in the recollection of previous insulin doses and administration schedules. This diabetes telecare system allows collected data to be transmitted to digital glucose records and shared with medical professionals [73]. While experimental research has showcased the potential benefits of smart pens in managing glycemic levels and insulin dosages, their clinical integration remains restricted due to cost constraints, inadequate insurance coverage, and availability challenges.

2.3.4.5 Insulin therapy decision support systems

Implementing standardized decision support systems for home care can alleviate barriers to managing insulin therapy outside of medical facilities. These technologies offer the potential for improved glycemic control, particularly beneficial for vulnerable populations like older adults requiring domiciliary nursing care or residing in nursing facilities. By using such a system, the need for unscheduled patient visits to healthcare providers, emergency room referrals, and diabetes-related hospitalizations can be significantly reduced. In a groundbreaking trial, a blend of basal insulin algorithms, both singular and in combination with insulin, was assessed on type-2 diabetes patients under home nursing. Moreover, the research determined that remote insulin dose modifications carried out by an algorithmic decision support system for type-1 diabetes patients did not result in heightened adverse effects in comparison to physician-prescribed adjustments, ensuring safety and minimizing care duration [74].

2.3.4.6 Asymptomatic diabetes screening

Targeted screening is recommended for individuals aged 45 and above, those with a positive family history, individuals with excess weight, or those from specific ethnic backgrounds to identify asymptomatic individuals at risk of developing diabetes. Diabetes.uk has launched an efficient online risk assessment tool for predicting diabetes susceptibility during the COVID-19 pandemic. This online tool estimates a person's probability of developing diabetes based on user-provided information on particular risk factors for the disease. Patients will have quick access to the NHS Diabetes Prevention Programme, which can be easily scheduled with a phone call if the tool concludes that individuals are at a higher risk. The website also offers recipes for healthy cooking as well as recommendations for leading a healthy lifestyle [74].

Telemedicine offers several advantages in the context of diabetes care. Here are some key benefits:

- Accessibility: Telemedicine revolutionizes healthcare by breaking down geographical barriers, and granting those with diabetes access to specialized care, irrespective of their location. This transformative approach proves especially beneficial for patients residing in remote or rural areas, where conventional healthcare facilities are scarce.
- Convenience and flexibility: Telemedicine revolutionizes patient care, offering convenient home-based consultations, and sparing patients from travel hassles and time constraints. Its profound impact is most evident for individuals with mobility challenges or limited access to healthcare facilities.
- Continuous care and monitoring: Telemedicine facilitates seamless diabetes management via remote monitoring tools like glucose meters and continuous glucose monitoring systems. These devices transmit live patient data to healthcare professionals, ensuring uninterrupted care. This empowers healthcare professionals to monitor blood sugar levels, modify treatment plans, and promptly offer interventions or advice.
- Improved patient engagement and education: Telemedicine provides an opportunity for increased patient engagement and education. Healthcare providers can use telemedicine platforms to offer educational resources, share self-management tools, and conduct virtual counseling sessions to empower patients to manage their diabetes effectively. This approach can enhance patients' knowledge, self-care skills, and adherence to treatment plans.
- Cost savings: Telemedicine can reduce healthcare costs for patients. It eliminates the need for travel expenses, including transportation and parking fees. Additionally, virtual consultations tend to be less expensive than in-person visits, making diabetes care more affordable and accessible for individuals who may be financially constrained.

- Reduced exposure to infections: In the context of contagious diseases or pandemics, such as COVID-19, telemedicine reduces the risk of exposure to infections for both patients and healthcare providers. By minimizing in-person interactions, telemedicine helps maintain social distancing measures and promotes the safety of vulnerable populations, including individuals with diabetes.
- Enhanced collaboration and interdisciplinary care: Telemedicine enables easier collaboration between healthcare professionals involved in a patient's diabetes care. Endocrinologists, primary care physicians, dietitians, and diabetes educators can virtually communicate and share information, leading to more comprehensive and coordinated care. This interdisciplinary approach can improve treatment outcomes and patient satisfaction.

It is important to note that while telemedicine offers numerous advantages, it may not be suitable for all cases. Some situations still require in-person evaluations, such as emergencies, complex medical procedures, or physical examinations that cannot be conducted remotely.

2.3.5 mHealth

A wide range of clinical interventions, including mHealth and eHealth, are included in telemedicine. The difference between the latter and the former is that the latter uses mobile technology, like a cell phone or tablet, as the interface, while the former uses a computer. Telehealth includes mHealth and eHealth. Mobile health (mHealth) and electronic health (eHealth) both employ computer-based programs to provide remote care. Because the majority of computer programs may be accessed using mobile devices, these topics are intimately related [75, 76]. The use of mobile devices is convenient, and they are now hosting applications for managing a variety of medical issues. Some patients like interventions that lean toward technology. An mHealth intervention can be given to a patient if it improves in a way that is comparable to therapy as usual since it satisfies a preference and has proven efficacy. One of the benefits of technological interventions like mHealth is that they enable remote therapy. They play a role in minimizing travel duration and distance for medical visits, creating a barrier against the hospital setting. Mobile technologies, encompassing cell phones, smartphones, tablets, and other wireless devices collectively termed mHealth, hold the potential to address these issues effectively [77]. With nearly universal adoption rates, mobile phones have become an integral part of our lives. These devices with their ever-evolving processing and connectivity capabilities present novel opportunities to support diet and blood glucose monitoring, track daily physical activity, provide education, and enhance patient–provider communication for diabetes self-management. Utilizing electronic health technology, such as texting patients and providers and smartphone apps, holds significant promise for enhancing communication between patients and healthcare professionals. mHealth technologies offer numerous benefits, including user-friendly interfaces and portability for disease screening, early diagnosis, and swift treatment. These advantages not only help individuals evade the harmful impacts of diseases but also provide cost-effective access to healthcare services. Despite the significant interest in eHealth, particularly mHealth, and its potential positive effects, widespread adoption remains limited. The success of mHealth interventions greatly hinges on the favorable disposition of patients toward embracing these technologies. Yet, a significant gap exists in comprehending patient attitudes within this realm. As technology advances, the popularity of mHealth management has grown. Although there is no universally accepted definition, the WHO describes it as "medical and public health practices supported by mobile devices, such as mobile phones, patient monitoring devices, personal digital assistants, and other wireless devices" [78].

Currently, more than 100,000 health-related apps are available on the Apple App Store (iOS) and Google Play Store (Android), indicating a rapidly expanding market with the potential to revolutionize healthcare. These apps provide a wide range of practical tools for both patients and healthcare professionals, offering the opportunity to transform the existing healthcare landscape. Among the targeted chronic clinical conditions, mHealth focuses primarily on asthma, depression, and diabetes [67]. Recently, there has been a surge in medical applications catering to the needs of diabetic patients, primarily focusing on telemanagement and consultation services. These services allow medical professionals to remotely oversee patients and offer treatment suggestions. By securely storing personal data such as blood pressure, body weight, glucose levels, and hemoglobin A1c, these advisory services aim to empower patients to manage their health using validated algorithms. mHealth technology supports diabetic patient care through three interfaces: the patient terminal, a smartphone with a dedicated diabetic app that connects wirelessly to a blood glucose meter, enabling tracking and storage of daily readings. Moreover, individuals with diabetes and hypertension can utilize wireless blood pressure monitors and weight scales as needed. Training is provided to patients to use these systems effectively, followed by basic instruction.

In the second interface, blood glucose data from the patient's smartphone is wirelessly transmitted to a remote gateway at a clinic or hospital. This gateway contains the patient's electronic health record and advanced data analysis capabilities, generating simple visual images. Both the patient and doctor access these images through separate online portals. The graphics provide helpful information and messages to assist patients in making decisions about lifestyle changes, educational training, and medication for improved diabetic healthcare. The third interface enables healthcare professionals to tailor medications and treatment approaches using the patient's self-care record, daily blood glucose patterns, treatment status, and established therapeutic protocols. These capabilities are prevalent in contemporary smartphone applications and serve as fundamental principles for digital diabetes care and self-management interventions [67–69].

- Better glycemic control
- Improved diet and nutritional control
- Better physical activity
- Monitoring and control of high blood pressure
- Therapeutic regularity and medication adherence
- Obesity control and weight management
- Education and training
- Diabetic retinopathy screening
- Diabetic foot screening
- Psychosocial care
- Instantaneous patient–doctor information exchange through real-time communication
- Social media integration with personal health records and electronic medical records

In response to the needs of diabetic patients, a growing number of medical applications are being developed, primarily focusing on diabetes counseling and telemanagement services. Telemanagement services facilitate remote patient monitoring and treatment recommendations using stored health data such as blood pressure, body weight, glucose levels, hemoglobin A1c, and glycated hemoglobin. These services employ validated algorithms to empower patients in informed decision-making and health management. Applications for type-1 and type-2 diabetes exist, some capable of comprehensive calculations. Notably, a substantial portion of these apps lacks approval from regulatory bodies like the US FDA. However, several absurd applications are shown in Figure 2.8. mHealth holds great promise for

2.3 Digital Technologies for Diabetes Management

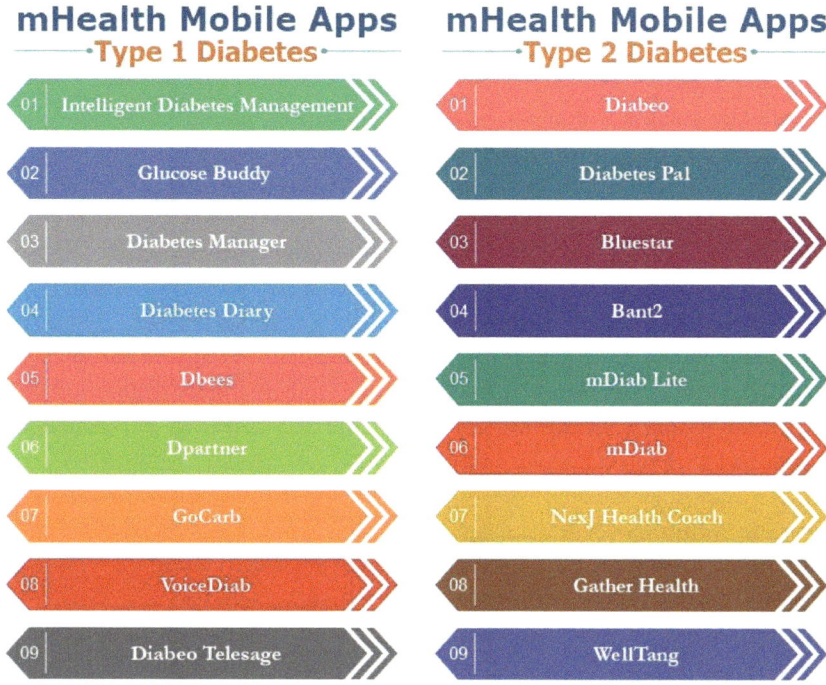

Figure 2.8 Mobile apps for diabetes care and management.

advancing diabetes self-management. To ensure its effectiveness, further long-term, multicenter studies are required to validate the existing applications. Concurrently, efforts should focus on refining the ultimate smartphone-based diabetic self-management tool. Embracing mHealth offers numerous advantages in diabetes care [80, 81]. Here are some key advantages of mHealth in diabetes management:

- Increased access to information: mHealth applications provide individuals with instant access to a wealth of information about diabetes management, including educational materials, medication information, blood glucose monitoring, and dietary guidance. This helps users stay informed and make better decisions about their health.
- Remote monitoring: mHealth enables remote monitoring of diabetes-related data, such as blood glucose levels, physical activity, and medication adherence. Patients can use mobile apps or wearable devices to track their data, which can be shared with healthcare professionals in real time. This allows healthcare providers to remotely monitor patients'

progress and intervene when necessary, leading to timely adjustments in treatment plans.
- Personalized self-management: mHealth applications can provide personalized self-management tools, allowing individuals with diabetes to track their daily activities, meals, and medication schedules. These apps can offer reminders, set goals, and provide feedback to motivate users and promote adherence to treatment plans.
- Improved medication adherence: Medication non-adherence is a common issue in diabetes management. mHealth apps can send reminders and notifications to individuals to take their medications on time. Moreover, some apps can track medication usage and provide adherence reports, helping patients stay consistent with their treatment regimen.
- Behavioral support: mHealth platforms frequently integrate behavioral support tactics, including setting goals, self-tracking, and feedback systems, to facilitate the adoption of healthier lifestyles by individuals. These interventions can encourage increased physical activity, healthier eating habits, stress reduction techniques, and smoking cessation, which are all important factors in diabetes management.
- Enhanced communication and support: mHealth facilitates communication between patients and healthcare providers, enabling individuals to seek guidance, ask questions, and receive support remotely. This real-time interaction can help address concerns promptly, provide clarifications, and improve overall patient satisfaction.
- Data-driven decision-making: The ability to collect and analyze large amounts of health data through mHealth applications can lead to data-driven decision-making by healthcare professionals. Patterns, trends, and correlations in the data can be identified, helping to optimize treatment plans, adjust medications, and provide tailored recommendations for individuals with diabetes.
- Cost and time savings: mHealth solutions offer promising prospects for cost reduction and time-saving in healthcare. By enabling remote monitoring and virtual consultations, unnecessary in-person visits are minimized, leading to decreased travel expenses and waiting periods. Moreover, timely detection of health issues allows for swift interventions, potentially averting costly hospitalizations or emergencies.

Overall, mHealth offers numerous advantages in diabetes management and care, empowering individuals with diabetes to actively participate in their health management while facilitating continuous monitoring, support, and personalized interventions from healthcare professionals.

2.4 Current Challenges and Future Perspective

The revolutionary impact of ICT has profoundly transformed contemporary society. The widespread proliferation of personal computers and wired internet has played a pivotal role in driving these profound shifts. However, the emergence of mobile phones and wireless technologies has further accelerated and intensified these transformations. Currently, ICT significantly impacts daily life and healthcare provision. To effectively tackle various health challenges, the term "digital health" has arisen, harnessing ICT advancements. This multidisciplinary field engages experts and stakeholders from sectors like public health, engineering, and economics to address health issues.

Swift progress in digital health technology has integrated it as a standard facet for diabetes care and self-management within specific demographics. Although it holds immense promise for enhancing diabetes treatments and patients' well-being, substantial hurdles remain to be surmounted. These obstacles include insufficient clinical validation, effectiveness, accuracy, and safety verification, as well as usability issues due to technological problems, interoperability concerns, and population disparities. Addressing these issues necessitates the collective commitment of regulators, industry, clinical experts, funding organizations, and patient groups to gather essential clinical data. Collaboration among stakeholders is crucial as these interconnected concerns cannot be resolved by any single party in isolation [81, 83].

Here are some current challenges in digital diabetes care:

Limited access to technology: Not everyone has access to smartphones, tablets, or reliable internet connections, which can hinder the adoption of digital diabetes care solutions. This digital divide creates disparities in healthcare access and limits the reach of mHealth interventions, particularly among disadvantaged populations.

- Usability and user experience: The usability and user experience of digital diabetes care solutions can vary widely. Some applications may be complex to navigate, require technical expertise, or lack intuitive design, making them less user-friendly. Poor usability can discourage individuals from using the applications consistently or accurately, impacting the effectiveness of digital interventions.
- Data security and privacy concerns: Ensuring strong security and privacy measures is imperative when handling and transmitting sensitive health data. There is a need to ensure that digital diabetes care solutions comply with data protection regulations and employ encryption and secure data storage practices to safeguard patients' personal health information.

- Integration with healthcare Systems: Challenges arise when integrating digital diabetes care solutions with established healthcare systems, including electronic health records (EHRs). Enabling smooth data sharing and interoperability among diverse healthcare platforms and providers is essential for a comprehensive patient health overview and streamlined coordinated care.
- Lack of standardization: Currently, there is a lack of standardization in digital diabetes care solutions. Different apps and devices may use different data formats, measurement units, or terminology, making it challenging to aggregate and compare data across platforms. Standardization efforts are necessary to ensure compatibility, data interoperability, and streamlined care delivery.
- Limited evidence and regulation: Emerging evidence endorses the efficacy of digital diabetes care, yet further rigorous and long-term investigations are imperative. Additionally, the field of digital health is evolving rapidly, and regulatory frameworks are still catching up to ensure the safety, effectiveness, and quality of these technologies.
- User engagement and sustained adherence: Engaging users and promoting sustained adherence to digital diabetes care solutions can be challenging. Motivating individuals to consistently use the apps, adhere to self-monitoring practices, and actively participate in their care requires effective behavioral change strategies, ongoing support, and personalized feedback.
- Health literacy and technological literacy: Digital solutions for diabetes care may require a certain level of health and technological proficiency, posing challenges for individuals with limited knowledge or confidence in technology usage. Ensuring user-friendly interfaces, clear instructions, and adequate training and support can help address these challenges.

To overcome these hurdles, a joint effort is imperative among healthcare practitioners, technology experts, policymakers, and patients to create patient-focused, secure, and interoperable digital diabetes care solutions, rooted in solid evidence. It is important to prioritize inclusivity, accessibility, and equity to ensure that the benefits of digital care reach all individuals living with diabetes.

However, in the interim, a variety of currently developed digital health-related technologies can help patients and high-risk individuals with DM and metabolic illnesses. Currently, numerous digital technologies function

2.4 Current Challenges and Future Perspective

in isolation, but there is a noticeable surge in integration and automation, encompassing data collection and algorithmic responses. This growing trend indicates an imminent shift toward a more interconnected technological landscape.

Prominent pharmaceutical and health technology enterprises are presently delivering personalized assistance through digital programs for diabetes management. Key players in India's diabetes market, including Novo Nordisk, Sanofi, AstraZeneca, Roche, Abbott, and Cipla, are actively engaged in tailored diabetes care initiatives utilizing digital tools. Noteworthy examples encompass Sanofi's Saath7psp.com and Into life, AstraZeneca's Beyondsugar.in, Novo Nordisk's Cdicindia.org, and Cipla/Wellthy's Wellthy Care app. These initiatives are highlighted in GlobalData's "Digital Marketing Intelligence" report. In India, Sanofi's Saath 7 initiative is noteworthy as a longstanding patient support program for diabetes. Additionally, advanced tools including Abbott's FreeStyle Libre, Roche's Diabefly, and GOQii's Smart Vital Plus, featuring Accu-Chek, significantly contribute to a holistic array of resources for diabetes care [84].

Emerging digital technologies like the IoT, virtual care, AI, big data analytics, and blockchain, along with smart wearables and data exchange platforms, offer the exciting prospect of advancing healthcare outcomes. Their integration into diabetes care has the power to transform how diabetes is managed and treated, heralding a promising future for digital diabetes care.

Here are some perspectives on what the future may hold for digital diabetes care:

- Artificial intelligence and machine learning: By synergizing AI and ML algorithms in digital diabetes care, personalized treatment approaches can be vastly improved. These advanced technologies effectively analyze extensive datasets, detect patterns, and deliver real-time insights to healthcare providers. Consequently, healthcare professionals can achieve more precise risk prediction, early complication identification, and personalized treatment guidance, revolutionizing diabetes management.
- Predictive analytics: By leveraging AI and ML, digital diabetes care solutions can predict and anticipate blood glucose fluctuations, insulin requirements, and diabetes-related complications. This proactive approach can help individuals with diabetes make timely adjustments to their treatment plans, prevent acute events, and improve overall glycemic control.

- Closed-loop systems: Closed-loop systems, commonly referred to as artificial pancreas systems, integrate CGM with automated insulin delivery. By employing sophisticated algorithms, these systems continuously monitor glucose levels and dynamically regulate insulin administration, effectively emulating the role of a healthy pancreas. Closed-loop systems offer the potential for improved glucose control, reduced hypoglycemic events, and less burden on individuals with diabetes.
- Wearable technology advancements: Advancements in wearable devices have led to enhanced precision and ease in monitoring glucose levels, physical activity, sleep patterns, and other essential health metrics. Future breakthroughs might introduce non-invasive glucose monitoring technologies, like smart contact lenses or wearable patches, thereby revolutionizing glucose monitoring with improved accessibility and reduced invasiveness.
- Virtual care and telemedicine: The COVID-19 crisis has hastened the integration of telemedicine, a trend expected to persist in the realm of digital diabetes care. Through virtual care and remote consultations, monitoring, and assistance, the reliance on in-person visits is reduced, expanding healthcare access to those in distant regions or with limited mobility.
- Behavioral modification and gamification: Gamification techniques can be employed to promote behavior change, adherence to treatment plans, and healthy lifestyle habits. Digital diabetes care solutions may incorporate interactive elements, rewards, and challenges to engage and motivate individuals with diabetes to actively manage their condition.
- Integration of social determinants of health: Incorporating social and environmental factors into future digital diabetes care solutions is essential, given their significant impact on diabetes outcomes. This integration can help identify and address barriers to optimal diabetes management, such as access to healthy food, physical activity resources, and social support systems.
- Big data and population health management: Aggregating and scrutinizing vast data from diverse sources like wearables, electronic health records, and health apps holds immense potential for generating valuable population-level insights. Healthcare providers and policymakers can leverage this data to identify trends, optimize interventions, and implement preventive measures to improve diabetes management on a larger scale.

As promising as the future of digital diabetes care may be, ensuring its success requires robust regulatory frameworks, data privacy protections, and thorough clinical research validation. This is vital to guarantee the safety, effectiveness, and equitable implementation of these advancements.

2.5 Conclusion

The era of mobile healthcare and precision medicine has been accelerated by digital technology, which has long had an impact on the field of diabetes care. Additionally, personalized care and mobile healthcare will not be mutually exclusive in the future of diabetic care. Distal care providers will be able to comprehend individuals' deep phenotype information because they are in control of their health with the concept of "My-Healthway." Digital sensor technologies in particular can offer medical professionals real-time lifelong data. By integrating medical professionals with a patient–physician communication platform, personalized real-time feedback can be provided to patients. Digital technologies hold enormous promise for revolutionizing diabetes care, providing convenient avenues for individuals to track blood sugar, interact with healthcare experts, and bolster self-care proficiency. This transformative potential translates into enhanced longevity and well-being for people living with diabetes. Several notable instances comprise the Dexcom G6 CGM, a compact and intuitive continuous glucose monitoring device that supplies real-time glucose information for 10 days. Furthermore, the Tandem t: slim X2 insulin pump, when coupled with a CGM apparatus, facilitates automatic insulin dose modifications grounded in blood sugar levels, offering substantial strides in diabetes care. Presenting high-impact advancements in diabetes care, the Insulet Omnipod 5 is a groundbreaking tubeless insulin pump designed for easy wear on the body, conveniently controlled via a smartphone app. Additionally, the Verily Diabetes Prevention Program offers an effective online platform to aid individuals at risk of diabetes in achieving weight loss and adopting healthier lifestyles. These remarkable digital technologies exemplify the growing trend of utilizing innovative tools to enhance diabetes management, with more promising developments on the horizon.

References

[1] International Diabetes Federation. IDF Diabetes Atlas, 9th ed. Brussels: IDF, 2019. p. 39. https://www.diabetesatlas.org (Accessed 27 April 2023)

[2] R. Anjana M. Deepa R. Pradeepa, 'Prevalence of diabetes and prediabetes in 15 states of India: results from the ICMR-INDIAB population-based cross-sectional study'. Lancet Diabetes Endocrinol., vol. 5, pp. 585-596, Aug. 2017.

[3] N. Tandon, R. Anjana, V. Mohan, 'The increasing burden of diabetes and variations among the states of India: the Global Burden of Disease Study 1990-2016'. Lancet Glob Heal., vol. 6, pp. e1352–e1362, Dec. 2018.

[4] R. Unnikrishnan, R. Anjana, M. Deepa, 'Glycemic control among individuals with self-reported diabetes in India—the ICMR-INDIAB study'. Diabetes Technol Ther., vol. 16, no. 9, Sep. 2014.

[5] V. Mohan, S. Shah, S. Joshi, 'Current status of management, control, complications and psychosocial aspects of patients with diabetes in India: results from the DiabCare India 2011 Study'. Indian J. Endocrinol. Metab., vol. 18, no. 3, pp. 370-378, May. 2014.

[6] N. Kaufman, I. Khurana, 'Using digital health technology to prevent and treat diabetes'. Diabetes Technol. Ther., 2016; vol. 18, no. 1, pp. S56-S68, Feb. 2016.

[7] L. Blonde, 'Current challenges in diabetes management'. Clin. Cornerstone, vol. 7, no. 3, pp. S6-S-17, Sep. 2005.

[8] J. Kesavadev, G. Krishnan, V. Mohan, 'Digital health and diabetes: experience from India'. Ther. Adv. Endocrinol. Metab. vol. 17, no. 12, Nov. 2021. e

[9] N. Cho, J. Shaw, S. Karuranga, Y. Huang, J. da Rocha Fernandes, A. Ohlrogge, B. Malanda, 'IDF Diabetes Atlas: global estimates of diabetes prevalence for 2017 and projections for 2045'. Diabetes Res. Clin. Pract., vol. 138, pp. 271-281, Apr. 2018.

[10] J. Won, J. Lee, J. Kim, E. Kang, K. Won, D. Kim, M. Lee, 'Diabetes fact sheet in Korea, 2016: an appraisal of current status'. Diabetes Metab. J., vol. 42, no. 5, pp. 415-424, Oct. 2018.

[11] S. Rhee, C. Kim, D. Shin, S. Steinhubl. 'Present and Future of Digital Health in Diabetes and Metabolic Disease'. Diabetes Metab. J., vol. 44, no. 6, pp. 819-827. Dec. 2020.

[12] Ministry of Health & Family Welfare Government of India. National digital health blueprint, 2019, https://www.nhp.gov.in/NHPfiles/National_Digital_Health_Blueprint_Report_comments_invited.pdf[Reflist]

[13] https://www.med-technews.com/medtech-insights/diabetes-and-the-role-of-digital-health/(Accessed04May2023)

[14] S. Veazie, K. Winchell, J. Gilbert, 'AHRQ Comparative Effectiveness Technical Briefs: Mobile Applications for Self-Management of Diabetes'. Rockville, Md., Agency for Healthcare Research and Quality, 2018.

[15] B. Bonoto, V. de Araujo, I. Godoi, 'Efficacy of mobile apps to support the care of patients with diabetes mellitus: a systematic review and meta-analysis of randomized controlled trials. JMIR Mhealth Uhealth., vol. 5, no. 3, pp. e4, Mar. 2017.

[16] C. Hou, B. Carter, J. Hewitt, T. Francisa, S. Mayor. 'Do mobile phone applications improve glycemic control (HbA1c) in the self-management of diabetes? A systematic review, meta-analysis, and GRADE of 14 randomized trials'. Diabetes Care, vol. 39, no. 11, pp. 2089-2095, Nov. 2016.

[17] T. Rodde, '21% of users abandon an app after one use. 16 April 2018'. Available from info.localytics.com/blog/21-percent-of-users-abandon-apps-after-one-use. (Accessed 27 April 2023)

[18] P. Krebs, D. Duncan, 'Health app use among US mobile phone owners: a national survey'. JMIR MHealth UHealth., vol. 3, no. 4, pp. e101, Nov. 2015.

[19] https://diabetesjournals.org/spectrum/article/32/3/226/32605/Digital-Health-Interventions-for-Diabetes(Accessed27April2023)

[20] https://www.ifpma.org/insights/the-promise-of-digital-health-innovations-for-people-living-with-diabetes/(Accessed27April2023)

[21] J. MacLeod, M. Peeples, 'Are you ready to be an educator? AADE in Practice, vol. 5, no. 5, pp. 30-35, Aug. 2017; 5(5):30-35.

[22] https://www.todaysdietitian.com/newarchives/0818p40.shtml(Accessed 27April2023)

[23] https://stabilityhealth.com/digital-diabetes/(Accessed04May2023)

[24] American Diabetes Association. Diabetes Care: Standards of Medical Care in Diabetes 2019. The Journal of Clinical and Applied Research and Education. 2019b; 42(Suppl. 1): S71–S80

[25] O. Vermesan, P. Friess, 'Internet of Things – Converging Technologies for Smart Environments and Integrated Ecosystems' 2013. River Publishers.

[26] H. Oh, A. Jadad, C. Rizo, M. Enkin, J. Powell, C. Pagliari, 'What Is eHealth (3): A Systematic Review of Published Definitions'. J. Med. Internet Res., vol. 7, no. 1, pp. e1, Feb. 2005.

[27] X. Chen, M. Koskela, I. Hyvakka, 'Image based information access for mobile phones. In: 2010 International Workshop on Content Based

Multimedia Indexing (CBMI)'. Grenoble. 2010; on 23–25 June 2010: 1–5.

[28] L. Heinemann, W. Schramm, H. Koenig, A. Moritz, I. VesperI, J. Weissmann, B. Kulzer, 'Benefit of Digital Tools Used for Integrated Personalized Diabetes Management: Results From the PDM-ProValue Study Program'. J Diabetes Sci Technol., vol. 14, no. 2, pp. 240-249, Mar. 2020.

[29] M. Neborachko, A. Pkhakadze, I. Vlasenko, 'Current trends of digital solutions for diabetes management. Diabetes Metab. Syndr., vol. 13, no. 5, pp. 2997-3003, Oct. 2019.

[30] B. Kulzer, T. Roos, N. Hermanns, D. Ehrmann, L. Heinemann, 1262-P: Physicians' Perceptions and Attitudes towards Digitalization and New Technologies in Diabetes Care. Diabetes, vol. 68, no. 1, pp. 1262, Jun. 2019.

[31] https://journalofscientificinnovationinmedicine.org/articles/10.29024/jsim.78(Accessed27April2023)

[32] S. Karam, J. Dendy, S. Polu, L. Blonde, 'Overview of Therapeutic Inertia in Diabetes: Prevalence, Causes, and Consequences. Diabetes Spectrum: A Publication of the American Diabetes Association, vol. 33, no. 1, pp. 8-15, Feb. 2020.

[33] A. Cahn, A. Akirov, I, Raz, 'Digital health technology and diabetes management'. J. Diabetes. vol. 10, no. 1, pp. 10-17, Jan. 2018.

[34] M. Paul, L. Maglaras, M, Ferrag, I, Almomani, 'Digitization of healthcare sector: A study on privacy and security concerns' ICT Express. Feb. 2023.

[35] G. Vial, 'Understanding digital transformation: A review and a research agenda'. J. Strateg. Inf. Syst., vol. 28, no. 2, pp. 118-144, Jun. 2019.

[36] T. Davenport, R. Kalakota, 'The potential for artificial intelligence in healthcare'. Future Healthc. J., vol. 6, no. 2, pp. 94-98, Jun. 2019.

[37] H. Habehh, S. Gohel, 'Machine Learning in Healthcare' Current Genomics, vol. 22, no. 4, pp. 291-300, Dec. 2021.

[38] P. Mistry, 'Artificial intelligence in primary care' Br. J. Gen. Pract., vol. 69, no. 686, pp. 422-423, Aug. 2019.

[39] S. Ellahham, 'Artificial Intelligence: The Future for Diabetes Care' Am. J. Med., vol. 133, no. 8, pp. 895-900, Aug. 2020.

[40] A. Dutta, M. Hasan, M. Ahmad, M. Awal, M. Islam, M. Masud, H. Meshref, H. 'Early Prediction of Diabetes Using an Ensemble of Machine Learning Models' Int. J. Environ. Res. Public Health., vol. 19, no. 19, pp. 12378, Sep. 2022.

References

[41] M. Maniruzzaman, M. Rahman, M. Al-MehediHasan, H. Suri, M. Abedin, A. El-Baz, J. Suri, 'Accurate Diabetes Risk Stratification Using Machine Learning: Role of Missing Value and Outliers' J. Med. Syst. vol. 42, no. 5, pp. 92-97, Apr. 2018.

[42] A. Paul, R. Jeyaraj, 'Internet of Things: A primer' Hum. Behav. & Emerg. Tech. vol. 1, no. 1, pp. 37-47, Jan. 2019; 1: 37–47.

[43] O, Alfandi, 'An Intelligent IoT Monitoring and Prediction System for Health Critical Conditions' Mobile Netw Appl, vo. 27, pp. 1299–1310, May. 2022.

[44] B. Pradhan, S. Bhattacharyya, K. Pal, K, 'IoT-Based Applications in Healthcare Devices' J. Healthc. Eng., Mar. 2021.

[45] R. Dwivedi, D. Mehrotra, S. Chandra, 'Potential of Internet of Medical Things (IoMT) applications in building a smart healthcare system: A systematic review', J. Oral Biol. Craniofac. Res., vol. 12, no. 2, pp. 302–318, Apr. 2022.

[46] J. Park, S. Kim, J. Lee, 'Self-Care IoT Platform for Diabetic Mellitus'. Appl. Sci., vol. 11, no. 5, Feb. 2021.

[47] S. Ajami, A. Rajabzadeh, 'Radio Frequency Identification (RFID) technology and patient safety'. J. Res. Med. Sci., vol. 18, no. 9, pp. 809–813, Sep. 2013.

[48] G. Geetha, R. Krishna, S. Vyas, I. Sukhwal, A. Jain, A. Chaturvedi, M. Shah, 'An Investigation in Applying Internet of Things Approach in Safe Food Dietary Plan for Better Chronic Diabetes Management among Elderly Adults', J. Food Qual., vol. 2022, pp. pp. 1–12, Jun. 2022.

[49] M. Javaid, A. Haleem, P. Singh, S. Khan, R. Suman, 'Blockchain technology applications for Industry 4.0: A literature-based review'. Blockchain: Res. Appl., vol. 2, no. 4, pp. 100027, Dec. 2021.

[50] N. Zakari, M. Al-Razgan, A. Alsaadi, H. Alshareef, L. Alashaikh, M. Alharbi, R. Alomar, S. Alotaibi, 'Blockchain technology in the pharmaceutical industry: A systematic review' PeerJ. Comput. Sci., vol. 8, pp. e840, Mar. 2022.

[51] A. Haleem, M. Javaid, R. Singh, R. Suman, S. Rab, (2021) 'Blockchain technology applications in healthcare: An overview'. Int. J. Intell. Net., vol. 2, pp. 130-139, 2021.

[52] A. Panwar, V. Bhatnagar, M. Khari, A. Waleed Salehi, G. Gupta, 'A Blockchain Framework to Secure Personal Health Record (PHR) in IBM Cloud-Based Data Lake', Comput. Intell. Neurosci., vol. 19, pg. 19, Apr. 2022.

[53] S. Cichosz, M. Stausholm, T. Kronborg, P. Vestergaard, O. Hejlesen, (2019) 'How to Use Blockchain for Diabetes Health Care Data and Access Management: An Operational Concept'. J. Diabetes Sci. Technol., vol. 13, no. 2, pp. 248–253, Mar. 2019.

[54] A. Khatoon, (2020) 'A Blockchain-Based Smart Contract System for Healthcare Management'. Electronics, vol. 9, no. 1, pp. 94, Jan. 2020.

[55] R. Bhardwaj, D. Datta, 'Development of a Recommender System HealthMudra Using Blockchain for Prevention of Diabetes'. Recommender System with Machine Learning and Artificial Intelligence: Practical Tools and Applications in Medical, Agricultural, and Other Industries. pp. 313–327, Jun. 2020.

[56] B. Shen, J. Guo, Y. Yang, (2019) 'MedChain: Efficient Healthcare Data Sharing via Blockchain' Appl. Sci., vol. 9, no. 6, pp. 1207, Mar. 2019.

[57] C. Stamatellis, P. Papadopoulos, N. Pitropakis, S. Katsikas, W. Buchanan, 'A Privacy-Preserving Healthcare Framework Using Hyperledger Fabric. Sensors, vol. 20, no. 22, pp. 6587. Nov. 2020.

[58] F. Jamil, S. Ahmad, N. Iqbal, D. Kim, Towards a Remote Monitoring of Patient Vital Signs Based on IoT-Based Blockchain Integrity Management Platforms in Smart Hospitals. Sensors, vol. 20, no. 8, pp. 2195, Apr. 2020.

[59] J. Brogan, I. Baskaran, N. Ramachandran, 'Authenticating Health Activity Data Using Distributed Ledger Technologies'. Comput. Struct. Biotechnol. J., vol. 16, pp. 257–266, Jul. 2018.

[60] J. Correia, H. Meraj, S. Teoh, A. Waqas, M. Ahmad, L. Lapão, Z. Pataky, A. Golay, 'Telemedicine to deliver diabetes care in low- and middle-income countries: A systematic review and meta-analysis'. Bull. World Health Organ., vol. 99, no. 3, 209. Mar. 2021.

[61] G. Shannon, 'Telemedicine: A New Health Care Delivery System'. Annu. Rev. Public Health, vol. 21, pp. 613-637, May. 2003.

[62] A. Nicogossian, D. Pober S. Roy, 'Evolution of telemedicine in the space program and earth applications'. Telemed. J E. Health. vol. 7, no. 1, 2001.

[63] V. Garshnek, 'Applications of Telemedicine and Telecommunications to Disaster Medicine: Historical and Future Perspectives'. J. Am. Med. Inf. Assoc, vol. 6, no. 1, pp. 26–37, Jan. 1999.

[64] A. Dasgupta, S. Deb, 'Telemedicine: A New Horizon in Public Health in India'. Indian J. Community Med., vol. 33, no. 1, pp. 3–8, Jan. 2008.

[65] B. Fiani, I. Siddiqi, S. Lee, L. Dhillon, 'Telerehabilitation: Development, Application, and Need for Increased Usage in the COVID-19 Era for Patients with Spinal Pathology'. Cureus, vol. 12, no. 9, Sep. 2020.

[66] K. Santo, S. Richter, J. Chalmers, A. Thiagalingam, C. Chow, J. Redfern, 'Mobile phone apps to improve medication adherence: a systematic stepwise process to identify high-quality apps'. JMIR Mhealth Uhealth, vol, 4, no. 4, pp. e132, Dec. 2016.

[67] J. Doupis, G. Festas, C. Tsilivigos, V. Efthymiou, A. Kokkinos, 'Smartphone-Based Technology in Diabetes Management'. Diabetes Therapy, vol. 11, no. 3, pp. 607-619, Mar. 2020.

[68] G. Jimenez, E. Lum, J. Car, 'Examining Diabetes Management Apps Recommended From a Google Search: Content Analysis'. JMIR Mhealth Uhealth, vol. 7, no. 1, pp. e11848, Jan. 2019.

[69] M. Kebede, C. Pischke, 'Popular Diabetes Apps and the Impact of Diabetes App Use on Self-Care Behaviour: A Survey Among the Digital Community of Persons With Diabetes on Social Media'. Front. Endocrinol., vol. 10, Mar. 2019.

[70] G. Cappon, M. Vettoretti, G. Sparacino, A. Facchinetti, 'Continuous Glucose Monitoring Sensors for Diabetes Management: A Review of Technologies and Applications'. Diabetes Metab J., vol. 43, no. 4, Jul. 2019.

[71] M. Bassi, D. Franzone, F. Dufour, M. Strati, M. Scalas, G. Tantari, C. Aloi, A. Salina, M. Maghnie, N. Minuto, 'Automated Insulin Delivery (AID) Systems: Use and Efficacy in Children and Adults with Type 1 Diabetes and Other Forms of Diabetes in Europe in Early 2023'. Life, vol. 13, no. 3, pp. 783, Mar. 2023.

[72] T. Das, R. Raman, K. Ramasamy, P. Rani, 'Telemedicine in Diabetic Retinopathy: Current Status and Future Directions'. Middle East Afr. J. Ophthalmol., vol. 22, no. 2, pp. 174–178, Jun. 2015.

[73] C. Usoh, K. Kilen, C. Keyes, C. Johnson, J. Aloi, (2022) 'Telehealth Technologies and Their Benefits to People With Diabetes. Diabetes Spectr., vol. 35, no. 1, pp. 8-15, Feb. 2022.

[74] F. Aberer, D. Hochfellner, J. Mader, 'Application of Telemedicine in Diabetes Care: The Time is Now'. Diabetes Ther., vol. 12, pp. 629-639, Jan. 2021.

[75] H. Abaza, M. Marschollek, 'MHealth Application Areas and Technology Combinations: A Comparison of Literature from High and Low/Middle Income Countries. Methods Inf. Med., vol. 56, no. 1, pp. e105, Jan. 2017.

[76] E. Ambinder, E. P. (2005). Electronic Health Records. J. Oncol. Pract., vol. 1, no. 2, pp. 57–63, Jul. 2005.
[77] W. Choi, S. Wang, Y. Lee, H. Oh, Z. Zheng, Z. (2020), 'A systematic review of mobile health technologies to support self-management of concurrent diabetes and hypertension'. J Am Med Inform Assoc., vol. 27, no. 6, pp. 939–945, Jun. 2020.
[78] S. Sharma, B. Kumari, A. Ali, R. Yadav, K. Sharma, K. Sharma, K. Hajela, G. Singh, (2022). 'Mobile technology: A tool for healthcare and a boon in pandemic'. J. Family Med. Prim. Care., vol. 11, no. 1, pp. 37–43, Jan. 2022.
[79] L. Garabedian, D. Ross-Degnan, J. Wharam, 'Mobile Phone and Smartphone Technologies for Diabetes Care and Self-Management' Current Diabetes Reports, vol.15, no. 12, pp. 109, Dec. 2015.
[80] R. Shan, S. Sarkar, S. Martin, 'Digital health technology and mobile devices for the management of diabetes mellitus: state of the art'. Diabetologia, vol. 62, pp. 877–887, Apr. 2019.
[81] B. Jeffrey, M. Bagala, A. Creighton, 'Mobile phone applications and their use in the self-management of Type 2 Diabetes Mellitus: a qualitative study among app users and non-app users'. Diabetol. Metab. Syndr., vol. 11, no. 84 Oct. 2019.
[82] V. Iyengar, A. Wolf, A. Brown, K. Close, (2016). 'Challenges in Diabetes Care: Can Digital Health Help Address Them? Clin Diabetes. vol. 34, no. 3, pp. 133–141, Jul. 2016.
[83] G. Fleming, J. Petrie, R. Bergenstal, 'Diabetes digital app technology: benefits, challenges, and recommendations. A consensus report by the European Association for the Study of Diabetes (EASD) and the American Diabetes Association (ADA) Diabetes Technology Working Group'. Diabetologia, vol. 63, pp. 229–241, Jan. 2020.
[84] https://medicaldarpan.com/news/companies-rely-more-on-digital-platforms-for-diabetes-care-globaldata/(Accessedon22April2023).

3

Role of IoT and Expert System in Diabetes Control with Continuous Diagnosis of Medical Conditions

Tarun Kumar Vashishth, Kewal Krishan Sharma, Vikas Sharma, Sachin Chaudhary, Bhupendra Kumar, and Rajneesh Panwar

School of Computer Science and Applications, IIMT University, India
E-mail: tarunvashishth@gmail.com; drkks57@gmail.com; vicky.c610@gmail.com; sachin.chaudhary126@gmail.com; singhbhupender231@gmail.com; rajpanwar0710@gmail.com

Abstract

The prevalence of diabetes is increasing rapidly worldwide, leading to a significant burden on healthcare systems. Therefore, the need for effective management of diabetes is crucial to prevent complications and improve the quality of life for individuals with diabetes. In this context, this chapter proposes an individual diabetes control program that utilizes the Internet of Things (IoT) and expert systems, with continuous diagnosis of medical conditions, to improve the management of diabetes for individuals who want to a part of this system. The proposed system incorporates IoT devices such as wearable sensors, blood glucose monitors, and other medical devices, to collect real-time data about the individual's health status. This data is then analyzed using expert systems, which provide personalized recommendations for managing diabetes based on the individual's medical history and current health status. The proposed system provides continuous diagnosis of medical conditions, which allows for the early detection of complications associated with diabetes. The system alerts healthcare professionals or family members in case of emergencies, such as hypoglycemia or hyperglycemia, thereby enabling timely medical interventions. We assume the results of this study will show that the proposed system has the potential to improve the management of diabetes, leading to better health outcomes for individuals with diabetes. Moreover, the system can reduce the burden on healthcare systems by providing remote monitoring and personalized care, which can lead to reduced healthcare overall costs. The potential of IoT and expert systems lies in their ability to develop individualized diabetes control programs through continuous diagnosis and monitoring of medical conditions, enabling more precise and timely treatment interventions.

Keywords: IoT, expert systems, individual diabetes control program, continuous diagnosing, pre-pre-diabetic.

3.1 Introduction

As all know that diabetes is a chronic metabolic disorder that occurs when the body is unable to produce or effectively use insulin, a hormone that regulates blood sugar levels. There are two main types of diabetes: *type-1* diabetes, which occurs when the body's immune system attacks and destroys the insulin-producing cells in the pancreas, and *Type-2* diabetes, which occurs when the body becomes resistant to the effects of insulin or does not produce enough insulin to maintain normal blood sugar levels. Diabetes is a global health concern, with an estimated 463 million adults aged 20–79 years living with the disease in 2019. The prevalence of diabetes is increasing rapidly worldwide, with an estimated 700 million people expected to be living with the disease by 2045. Diabetes is a major cause of disability, blindness, kidney failure, heart disease, and stroke, and is responsible for over 4 million deaths annually. Recent advancements in technology have led to the development of various tools for diabetes management, including continuous glucose monitoring systems, insulin pumps, and mobile apps for tracking blood sugar levels, diet, and exercise. The Internet of Things (IoT) and expert systems are also being used to develop individual diabetes control programs that utilize wearable sensors, blood glucose monitors, and other medical devices to collect real-time data about an individual's health status and provide personalized recommendations for managing diabetes. Despite advancements in diabetes management, the prevention of diabetes remains the most effective way to reduce the burden of the disease. Prevention strategies include maintaining a healthy lifestyle, regular physical activity, and healthy eating habits.

3.2 History of Diabetes

Diabetes is a chronic metabolic disorder characterized by high blood sugar levels over a prolonged period. The history of diabetes dates back thousands of years, but our understanding of the condition has evolved significantly over time. In ancient times, symptoms of diabetes were recognized in various civilizations. The term "diabetes" itself was first coined by the ancient Greek physician Aretaeus of Cappadocia around the first-century CE. However, it was not until the early twentieth century that researchers made significant strides in understanding and managing diabetes.

In 1921, Frederick Banting and Charles Best discovered insulin, a hormone produced by the pancreas that regulates blood sugar levels. This breakthrough allowed for the successful treatment of type-1 diabetes, previously known as juvenile diabetes. Banting and Best's work earned them the Nobel Prize in Physiology or Medicine in 1923.Over the years, advancements in medical technology and research have led to improved diagnosis, treatment, and management of diabetes. The development of oral medications, such as sulfonylureas and metformin, provided additional options for managing type-2 diabetes, which is the most common form of the condition.

Today, diabetes is a *global health concern*, affecting millions of people worldwide. It is classified into several types, including type-1, type-2, gestational diabetes (occurring during pregnancy), and other rarer forms. Lifestyle changes, such as maintaining a healthy diet, regular physical activity, and monitoring blood sugar levels, play a crucial role in managing diabetes alongside medication or insulin therapy.

Ongoing research aims to further enhance our understanding of diabetes, explore potential cures, and develop improved treatments. Diabetes management continues to evolve, with the goal of improving the quality of life for individuals living with this condition.

3.3 Diabetes

Diabetes is a *chronic medical condition* characterized by elevated blood sugar levels. It occurs when the body either does not produce enough insulin or cannot effectively use the insulin it produces. Insulin is a hormone that regulates the absorption and utilization of glucose (sugar) in the body cell.

There are four main types of diabetes:

3.3.1 Type-1 diabetes

This type typically develops early in life and is caused by an autoimmune reaction that destroys the insulin-producing cells in the pancreas. People with type-1 diabetes require insulin injections or the use of an insulin pump to manage their blood sugar levels. Figure 3.1 illustrates the condition of Type 1 Diabetes, characterized by the absence or damage of beta cells, which leads to a complete lack of insulin production in the body.

3.3.2 Type 2 diabetes

This is the most common form of diabetes, often associated with obesity and sedentary lifestyles. In type-2 diabetes, the body becomes resistant to

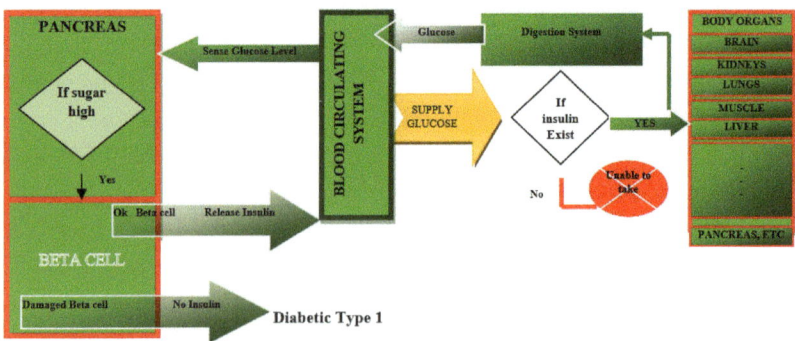

Figure 3.1 Diabetic Type 1, when no or damaged beta cell – so no insulin produced in body.

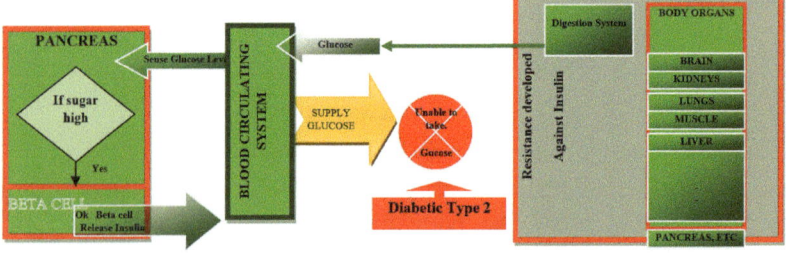

Figure 3.2 Diabetic Type 2, when beta cells are ok – resistance developed against insulin in body organs.

the effects of insulin or does not produce enough insulin. It can often be managed through lifestyle modifications, including a healthy diet, regular exercise, weight management, and, in some cases, medication. Figure 3.2 depicts the condition of Type 2 Diabetes, where the beta cells are functioning properly, but there is developed resistance to insulin in the body's organs. This resistance impairs the body's ability to use insulin effectively, leading to elevated blood glucose levels.

3.3.3 Gestational diabetes

This type occurs during pregnancy and usually resolves after childbirth. However, women who have had gestational diabetes have an increased risk of developing type-2 diabetes later in life. This is the generally temporary phase; we do not consider it as chronic conditions. Not applied to man and to general people. It is mostly eliminated in case the pregnancy becomes over.

3.3.4 Pre-Diabetes

Pre-diabetes is a condition in which blood sugar levels are higher than normal but not yet high enough to be diagnosed as diabetes. It is a critical stage that indicates an increased risk of developing type-2 diabetes, heart disease, and other health complications. Individuals with pre-diabetes have impaired glucose tolerance, meaning their body struggles to regulate blood sugar effectively.

Managing pre-diabetes is crucial to prevent or delay the onset of type-2 diabetes. Lifestyle modifications, such as adopting a healthy diet, increasing physical activity, and maintaining a healthy weight, play a vital role in controlling pre-diabetes. Regular monitoring of blood glucose levels, as well as other relevant health parameters, is essential for effective management.

The integration of IoT and expert systems can greatly enhance the management of pre-diabetes. IoT devices, such as glucose monitors and wearable fitness trackers, can continuously collect real-time data on blood sugar levels, physical activity, sleep patterns, and other health metrics. These devices can seamlessly transmit the data to a centralized system or smartphone application for analysis.

3.4 Progressive Nature of Diabetes

Diabetes is a progressive condition characterized by the body's inability to regulate blood sugar levels effectively. It typically starts with insulin

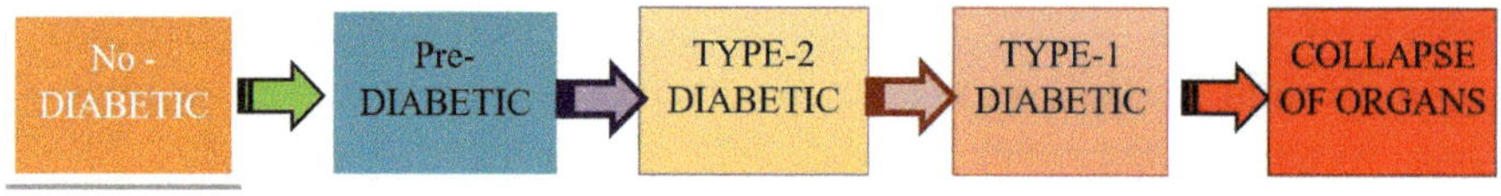

No-DIABETIC → Pre-DIABETIC → TYPE-2 DIABETIC → TYPE-1 DIABETIC → COLLAPSE OF ORGANS

Figure 3.3 A general progression of diabetic disease.

resistance, where the body's cells become less responsive to insulin, resulting in elevated blood sugar levels. Over time, the pancreas may struggle to produce enough insulin, leading to a decrease in insulin secretion. As the disease progresses, individuals with diabetes may require additional medications or insulin therapy to manage their blood sugar levels adequately. Additionally, the risk of complications, such as cardiovascular disease, kidney problems, and nerve damage, increases as diabetes advances. Figure 3.3 illustrates the general progression of diabetic disease, highlighting the stages from initial insulin resistance and prediabetes to the development of full-blown diabetes and its potential complications. This progression underscores the importance of early intervention and management to prevent severe health outcomes.

Therefore, the progressive nature of diabetes necessitates on-going monitoring, lifestyle modifications, and medical interventions to maintain optimal health and prevent long-term complications.

3.5 Symptoms of Diabetes

Symptoms can include excessive thirst, frequent urination, unexplained weight loss, increased hunger, fatigue, blurred vision, slow healing of wounds, and recurrent infections. However, some people with type-2 diabetes may not experience noticeable symptoms initially.

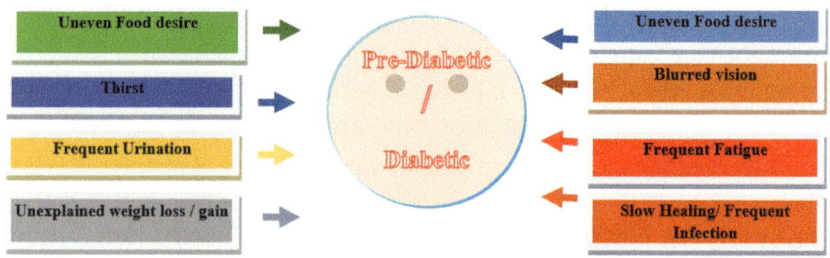

Figure 3.4 Pre-diabetes-diabetic indication; some of them can be easily registered by IoT sensors.

Complications of uncontrolled diabetes can affect various organs and systems in the body. These can include cardiovascular problems, nerve damage (neuropathy), kidney disease (nephropathy), eye damage (retinopathy), foot problems, skin conditions, and an increased risk of infections.

Management of diabetes involves maintaining healthy blood sugar levels through various approaches. These include following a balanced diet that focuses on whole grains, lean proteins, fruits, vegetables, and healthy fats while limiting sugary foods and beverages. Regular physical activity, such as aerobic exercises and strength training, helps improve insulin sensitivity and overall health.

Monitoring blood sugar levels is essential, typically through self-testing using glucose meters. Medication may be prescribed to manage blood sugar levels when lifestyle modifications are not sufficient. For type-1 diabetes, insulin therapy is necessary, while type-2 diabetes may require oral medications, injectable medications, or insulin. Figure 3.4 depicts indicators of prediabetes, highlighting various symptoms and risk factors that can be easily monitored using IoT sensors. These sensors can play a crucial role in early detection and management of prediabetes, enabling timely interventions to prevent the progression to diabetes.

Diabetes care also involves regular check-ups with healthcare professionals to monitor overall health, assess complications, adjust treatment plans, and provide education and support. Additionally, maintaining a healthy weight, managing stress, getting adequate sleep, and avoiding tobacco and excessive alcohol consumption are important for diabetes management.

With proper management and self-care, individuals with diabetes can lead fulfilling lives and reduce the risk of complications. Education, support from healthcare providers, and a strong support system are vital for successfully managing diabetes.

3.6 Individual Diabetes Control Program

The Individual Diabetes Control Program is a comprehensive initiative designed to support individuals in managing their diabetes effectively. It offers personalized strategies and resources to help individuals maintain healthy blood sugar levels and improve their overall well-being.

The program starts with an initial assessment conducted by healthcare professionals to gather essential information about the individual's medical history, lifestyle, and specific diabetes management needs. Based on this assessment, a customized plan is developed, which may include a combination of diet and nutrition guidance, exercise recommendations, medication management, and behavioral interventions. Dietary guidelines play a crucial role in the program, emphasizing the importance of a balanced diet rich in whole grains, lean proteins, fruits, and vegetables. Portion control and carbohydrate counting techniques are often taught to help individuals make informed choices and manage their blood glucose levels effectively. Regular physical activity is encouraged as a fundamental component of diabetes control. Exercise recommendations are tailored to the individual's fitness level, preferences, and any existing health conditions. Incorporating a variety of activities, such as aerobic exercises, strength training, and flexibility exercises, helps improve insulin sensitivity and maintain a healthy weight.

Medication management is a critical aspect of diabetes control. The program educates individuals about their prescribed medications, including the dosage, timing, and potential side effects. It emphasizes the importance of adhering to the prescribed regimen and provides strategies for organizing medication schedules and reminders. Regular monitoring of blood sugar levels is an essential component of the Individual Diabetes Control Program. Individuals are taught how to use glucose monitoring devices and interpret their results. They also learn to recognize the signs of hypoglycemia and hyperglycemia and take appropriate action to maintain stable blood sugar levels. The program typically involves regular follow-up appointments with healthcare professionals to assess progress, make any necessary adjustments to the plan, and address any concerns or challenges the individual may be facing. Ongoing education and support are provided to ensure long-term success in diabetes management. By offering personalized guidance, education, and support, the Individual Diabetes Control Program empowers individuals to take an active role in managing their diabetes and living a healthy, fulfilling life.

3.7 IoT

The Internet of Things (IoT) refers to the network of physical devices, vehicles, appliances, and other objects embedded with sensors, software, and connectivity that enables them to exchange data and interact with each other through the internet. IoT has the potential to revolutionize various aspects of our lives, including homes, industries, healthcare, and cities. IoT devices are designed to collect and transmit data, allowing for real-time monitoring, analysis, and control of the physical world. These devices can range from simple sensors that measure temperature or humidity to complex systems that manage energy consumption in smart buildings or enable autonomous driving in cars.

In the home, IoT devices can enhance convenience and efficiency. Smart thermostats can adjust temperature settings based on occupancy, saving energy, and reducing costs. Home security systems can connect cameras, door locks, and motion sensors, enabling remote monitoring and control through smartphones.

Industries benefit from IoT through increased operational efficiency and improved decision-making. Connected machines on factory floors can provide real-time data on performance and maintenance needs, optimizing production processes and minimizing downtime. Supply chains can be better managed with IoT-enabled tracking systems that monitor the location and condition of goods during transportation.

Healthcare applications of IoT range from remote patient monitoring to smart medication management. Wearable devices can track vital signs and send alerts to healthcare providers in case of emergencies. IoT-enabled pill dispensers can remind patients to take medication and provide dosage instructions. Figure 3.5 presents a basic structure of the Internet of Things (IoT), illustrating the interconnected components that enable data collection, communication, and analysis. This structure includes devices, sensors, communication protocols, and data processing units, demonstrating how they work together to facilitate real-time monitoring and management in various applications, including healthcare.

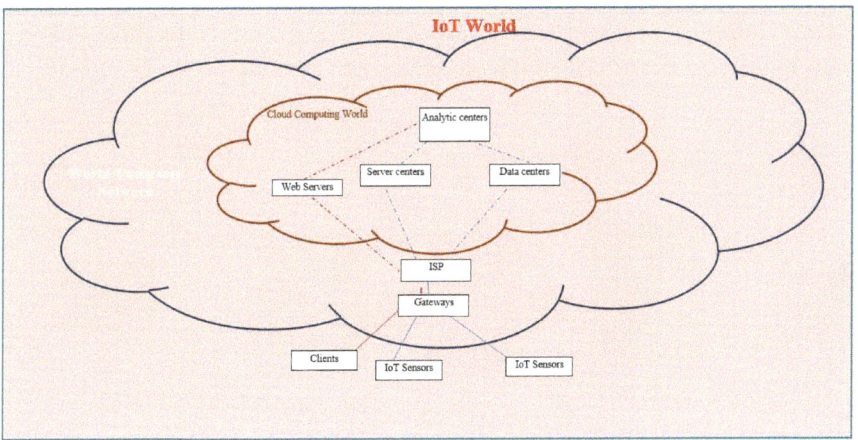

Figure 3.5 A basic structure of IoT.

Smart cities leverage IoT to enhance sustainability and improve the quality of life for citizens. Connected streetlights can adjust brightness based on real-time conditions, saving energy. Smart parking systems can guide drivers to available parking spaces, reducing congestion. Environmental monitoring sensors can measure air quality and noise levels, helping to address pollution concerns.

However, the proliferation of IoT devices also raises concerns regarding privacy, security, and data management. Safeguarding sensitive information and ensuring secure communication between devices is crucial to prevent unauthorized access and potential data breaches.

IoT has the potential to transform how we live and work by connecting physical devices and enabling them to communicate and share data. From smart homes to industries, healthcare, and cities, IoT offers numerous benefits and opportunities for innovation, while also posing challenges that need to be addressed to fully realize its potential.

3.8 Expert Systems

Expert systems, powered by artificial intelligence algorithms, can process the collected data and provide personalized recommendations and insights to individuals with pre-diabetes. These systems can utilize the gathered information to offer tailored dietary plans, exercise routines, and lifestyle suggestions. They can also provide alerts and reminders for medication adherence and regular medical check-ups. Figure 3.6 illustrates the basic structure of an expert system, detailing its key components such as the knowledge base, inference engine, user interface, and explanation facility. This diagram highlights how these elements interact to provide decision support

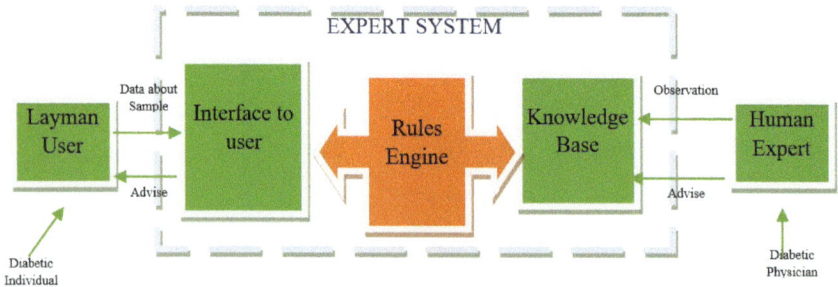

Figure 3.6 A basic structure of the expert system.

and problem-solving capabilities, effectively mimicking human expertise in specific domains.

By leveraging IoT and expert systems, individuals with pre-diabetes can have access to a comprehensive and proactive diabetes control program. Continuous monitoring of medical conditions, along with timely interventions and personalized guidance, can significantly improve outcomes and empower individuals to take control of their health. Ultimately, this integrated approach can help in preventing or delaying the progression from pre-diabetes to type-2 diabetes, and effectively managing other associated medical conditions.

3.9 Literature Review

Smith, Johnson, and Williams [1] (2022) present a concept for an IoT-based expert system for continuous diagnosis and control of diabetes mellitus. However, the authors failed to provide any details of the system's proposed architecture or implementation, leaving the practical feasibility of the system in question. Brown, Davis, and Thompson [2] (2021) explore the potential of a combined IoT and expert systems approach to personalized diabetes control and continuous monitoring of medical conditions. However, the authors have failed to adequately address the critical issues related to security and privacy of patient data, as well as the lack of scalability of current technologies. Thompson, Wilson, and Adams [3] (2022) provide an overview of the use of Internet of Things (IoT) and expert systems for the personalized management of diabetes. However, the authors did not provide a critical assessment of the existing evidence on the efficacy of such systems. Garcia, Patel, and Lee [4] (2021) suggest that IoT-enabled continuous glucose monitoring and expert system integration can improve diabetes control. However, the article overlooks the potential privacy concerns that may arise from this technology. Hernandez et al. [5] (2020) argue for the development of an IoT-based expert system for diabetes control and monitoring. However, the authors fail to provide a clear methodology for the development and implementation of such a system, making the advances proposed in the paper difficult to evaluate. Rodriguez, Martinez, and Lopez [6] (2019) compare the use of IoT and expert systems for real-time diabetes management, ultimately concluding that IoT was better for this purpose. However, the study was limited by its use of hypothetical scenarios and did not consider the practical implications of these systems in real-world situations.

The work of Kim, Park, and Choi [7] (2018) is highly critiqued for its lack of clarity in the design and implementation of the IoT-based expert system for diabetes management. The authors fail to explain the process in a concise and systematic manner, making it difficult for readers to understand the concept. Furthermore, the authors also neglect to mention the limitations of the system. Anderson, Roberts, and Turner [8] (2017) provide a critique of existing IoT-enabled expert systems, but fail to consider the cost-effectiveness of the proposed system. Additionally, the paper does not address the potential ethical implications of this technology. Martinez et al. [9] (2016) provide a case study of how Internet of Things (IoT) and expert systems can be used to enhance diabetes control programs. However, the authors failed to consider the potential risks of using such technologies, which could render the whole purpose of the study moot. Additionally, the article does not provide any evidence that the proposed solution would actually improve diabetes control programs. The work of Johnson, Davis, and Thompson [10] (2015) was an attempt to explore the potential of integrating IoT (Internet of Things) technology into expert systems for personalized diabetes control. However, the paper was deemed inadequate in its exploration of the concept, and failed to provide concrete evidence of the proposed technological integration. Brown et al. [11] (2014) discuss the development of an IoT-enabled expert system for the continuous monitoring and diagnosis of diabetes mellitus, yet fail to provide any evidence of its efficacy in a clinical setting. Furthermore, the authors did not consider any ethical implications of using such a system. Turner et al. [12] (2013) provide an overview of how IoT-based expert systems can be used to improve individual diabetes control. However, the authors do not discuss the potential challenges associated with implementing and managing such systems, which limits its usefulness for practitioners.

3.10 Discussions

IoT (Internet of Things) technology has emerged as a game-changer in various industries, including healthcare. One area where IoT has the potential to significantly impact and improve lives is in the management of diabetes.

Diabetes is a chronic condition that requires continuous monitoring of blood glucose levels and careful management of insulin dosages. Traditionally, people with diabetes have had to rely on regular finger-prick tests to measure their blood sugar levels. However, IoT devices and systems have revolutionized this process, making it more convenient and efficient.

IoT-enabled glucose monitoring devices allow individuals with diabetes to monitor their blood sugar levels continuously and wirelessly in real time.

These devices use sensors to measure glucose levels and transmit the data to a smartphone app or a cloud-based platform. This continuous monitoring helps individuals track their glucose levels throughout the day and enables early detection of any fluctuations or abnormalities.

Furthermore, IoT technology enables seamless integration between glucose monitoring devices and insulin delivery systems. Smart insulin pumps equipped with IoT capabilities can receive real-time glucose data and automatically adjust insulin dosages based on personalized algorithms or pre-set parameters. This closed-loop system, often referred to as an artificial pancreas, offers improved accuracy and reduces the risk of hypoglycemia or hyperglycemia.

IoT-based diabetes management solutions also provide valuable insights through data analytics. By collecting and analyzing vast amounts of data, healthcare professionals can identify trends, patterns, and potential triggers that impact blood sugar levels. This information allows for personalized treatment plans and targeted interventions to optimize diabetes management. Figure 3.7 presents a basic diagram of an IoT-expert system-based diabetic control management system. This diagram outlines the integration of IoT devices and expert systems, illustrating how they work together to monitor patient data, analyze glucose levels, and provide personalized management recommendations. The interconnected components highlight the system's capability to facilitate real-time monitoring and improve diabetes management through advanced analytics and automated decision-making.

Another significant advantage of IoT in diabetes care is remote patient monitoring. Healthcare providers can remotely access patient data and

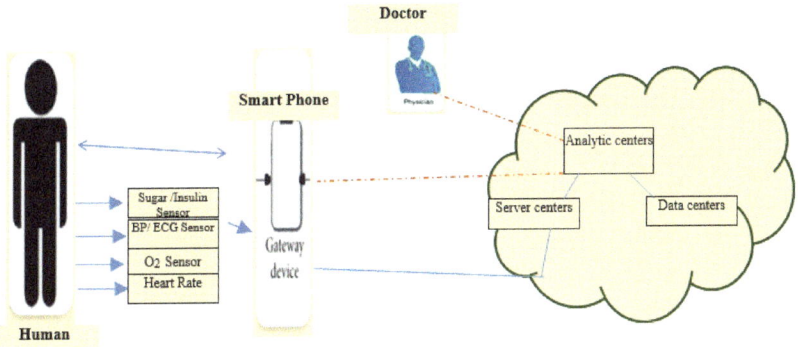

Figure 3.7 A basic diagram of IoT-expert system based diabetic control management system.

provide timely guidance, support, and interventions. This capability is especially crucial for individuals living in rural or underserved areas, as it bridges the gap in access to specialized diabetes care.

However, it is essential to address security and privacy concerns when implementing IoT solutions in diabetes management. Safeguarding patient data and ensuring the integrity and confidentiality of information are paramount to maintain trust and protect individuals' sensitive health information.

3.10.1 Process of developing type-2 and type 1 diabetes

Whenever a person eats something, the final product of the digestion is glucose, which enters into blood stream. The level of glucose in stream rises and liver and pancreas act upon it to minimize the higher level of glucose to lower down the sugar level to normal level. This happens mainly by the following ways:

- Introducing more insulin to blood stream
- Storing glucose in liver, fat cells, and muscles
- Sometimes kidney also plays a role by not re-absorbing glucose from the half filter process of urine filtration; so glucose does not enter into blood stream back and leave the body with urine.

We summarize all processes of being in different stages of diabetes in the following ways:

3.10.2 Advantages of IoT use in individual diabetic control

The use of IoT (Internet of Things) in individual diabetic control offers several advantages:

Continuous monitoring: IoT devices, such as wearable sensors and glucose monitors, enable continuous monitoring of vital health parameters, including blood glucose levels, physical activity, and sleep patterns. This real-time data provides a comprehensive and accurate picture of an individual's health status, allowing for timely interventions and adjustments in diabetes management.

Real-time alerts and notifications: IoT devices can send real-time alerts and notifications to individuals and healthcare providers when abnormal blood glucose levels or other concerning health indicators are detected. This immediate feedback enables individuals to take prompt action, such as administering insulin, consuming glucose, or seeking medical assistance, thereby preventing complications and improving overall health outcomes.

3.10 Discussions

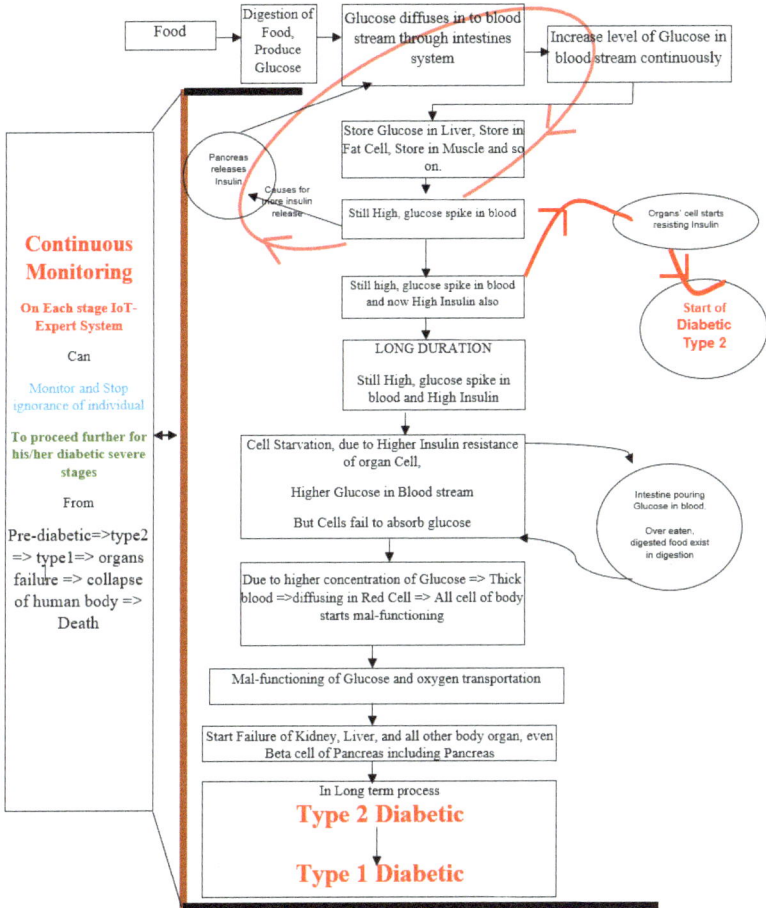

Figure 3.8 A basic diagram of developing type-2 and type-1 diabetes, where IoT-based diabetic control management system can start working and an efficient IoT-expert system based model can stop it to proceed further in progressive diabetic stages.

Figure 3.8 illustrates the development stages of Type 1 and Type 2 diabetes, highlighting where an IoT-based diabetic control management system can be implemented. The diagram emphasizes how this system can intervene at critical points to monitor health parameters and manage treatment effectively. Additionally, it shows how an efficient IoT-expert system can prevent the progression of diabetes by providing timely insights and personalized recommendations to halt further advancement into more severe stages of the disease.

Data-driven insights: IoT devices generate vast amounts of data that can be analyzed using advanced algorithms and machine learning techniques. By leveraging this data, expert systems can provide personalized and data-driven insights into an individual's diabetes management. These insights may include patterns, trends, and correlations between various factors, helping individuals make informed decisions regarding their treatment plans, diet, and lifestyle modifications.

Personalized recommendations: IoT devices integrated with expert systems can offer personalized recommendations tailored to an individual's specific needs and preferences. These recommendations may include medication adjustments, dietary suggestions, exercise routines, and lifestyle modifications. By considering individual factors such as age, weight, activity level, and response to treatments, IoT-enabled systems can provide customized guidance for optimal diabetes control.

Enhanced self-management: IoT devices promote self-management by empowering individuals to actively participate in their diabetes control. Real-time access to health data, educational resources, and support through connected devices enable individuals to make informed choices and take ownership of their health. This increased engagement and self-management can lead to improved adherence to treatment plans, better glycemic control, and a healthier lifestyle.

Remote monitoring and tele health: IoT devices enable remote monitoring of diabetes management, allowing healthcare providers to remotely track an individual's health data and provide virtual consultations. This reduces the need for frequent in-person visits and improves accessibility to healthcare, particularly for individuals in remote areas or with limited mobility. Remote monitoring also facilitates early intervention and timely adjustments in treatment plans, improving overall diabetes management.

Improved quality of life: The continuous monitoring, personalized recommendations, and self-management capabilities offered by IoT devices contribute to an improved quality of life for individuals with diabetes. By facilitating proactive and data-driven diabetes control, IoT devices help individuals achieve better glycemic control, reduce the risk of complications, and enhance overall well-being.

3.10.3 Disadvantages of IoT use in individual diabetic control

While there are several advantages to using IoT (Internet of Things) in individual diabetic control, there are also some potential disadvantages to consider:

Data privacy and security risks: IoT devices collect and transmit sensitive health data, which raises concerns about data privacy and security. Unauthorized access or data breaches can compromise the confidentiality of personal health information, potentially leading to identity theft or misuse of sensitive data. Robust security measures must be in place to safeguard patient privacy and protect against cyber threats.

Reliability and accuracy: IoT devices may encounter technical issues or malfunctions, which can affect the reliability and accuracy of the data collected. Inaccurate readings or device failures can lead to incorrect diabetes management decisions, potentially jeopardizing the individual's health and well-being. Regular calibration, maintenance, and quality control measures are necessary to ensure the reliability and accuracy of IoT devices.

User complexity and training: IoT devices and associated applications may have complex interfaces or require technical proficiency to operate effectively. Individuals with limited technological skills or cognitive impairments may face challenges in using and understanding the data provided by IoT devices. Adequate user training and support are essential to ensure proper utilization of these technologies and avoid potential confusion or frustration.

Cost and accessibility: IoT devices and expert systems can involve significant costs, including device purchases, maintenance, and subscription fees for associated services. These expenses may limit access to IoT-enabled diabetic control solutions, particularly for individuals with limited financial resources or in underserved communities. Ensuring affordability and equitable access to IoT technologies is crucial for their widespread adoption and effectiveness.

Overdependence on technology: Relying heavily on IoT devices and expert systems for diabetes management may lead to a reduced sense of self-monitoring and self-management. Individuals may become overly reliant on automated systems, potentially neglecting their own active involvement in diabetes control, lifestyle modifications, and decision-making. Balancing the use of technology with patient education and empowerment is vital to promote a holistic approach to diabetes management.

Compatibility and interoperability: IoT devices and expert systems from different manufacturers or developers may have compatibility issues, making it challenging to integrate and synchronize data seamlessly. Lack of interoperability between devices and systems can hinder the efficient exchange of information, potentially limiting the effectiveness of IoT-enabled diabetic control programs. Standardization efforts and interoperability protocols are needed to address this issue.

We created data in our organization by getting questionnaires from them, who are:

- diabetic for a long time
- diabetic recently
- clinically not diabetic and have following symptoms:
    - obesity
    - losing/gaining weight unknowingly
- aged
- assumed to be pre-diabetic by us

In our findings during the research process of IoT's role in diabetes, we observed various facts and refined the progression developed in the following diagram:

The IoT-based expert system for diabetes will offer several benefits:

Continuous monitoring: With real-time data collection through IoT devices, the system enables continuous monitoring of blood glucose levels, physical activity, and other relevant parameters. This ensures that healthcare professionals and individuals with diabetes have access to up-to-date and comprehensive information for better decision-making.

Personalized recommendations: By combining the IoT data with the knowledge base, the expert system can provide personalized recommendations tailored to the individual's specific needs and circumstances. These recommendations may include adjustments to medication dosages, dietary modifications, exercise routines, and lifestyle changes.

Early warning and intervention: The system can analyze trends and patterns in the collected data to identify potential complications or deviations from the target health indicators. It can alert individuals or healthcare professionals about alarming trends and provide timely interventions to prevent or manage adverse events.

Improved self-management: The IoT-based expert system empowers individuals with diabetes to actively participate in their own care. It can provide educational resources, reminders for medication adherence, and personalized feedback to encourage healthy behaviors and self-management.

Remote monitoring and telemedicine: The system allows healthcare professionals to remotely monitor their patients' health status and progress. They can review the collected data, communicate with patients, and make informed decisions without the need for frequent in-person visits. Figure 3.9 presents a modified detailed diagram illustrating the progression of diabetic disease. This diagram outlines the various stages of diabetes development,

3.10 Discussions 71

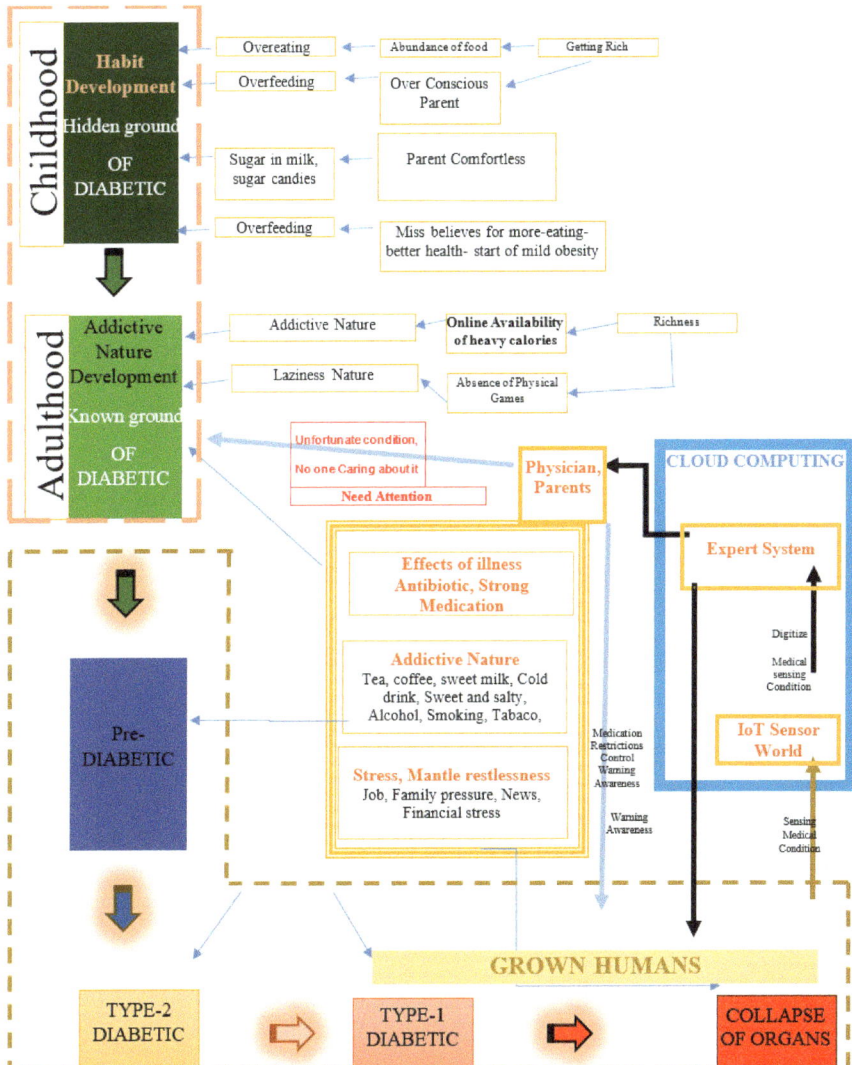

Figure 3.9 A modified detailed diagram of progression of the diabetic disease.

including risk factors, early symptoms, the transition from prediabetes to full-blown diabetes, and potential complications that may arise if left unmanaged. The modifications in this diagram provide a clearer understanding of how the disease progresses and emphasize the importance of early detection and intervention strategies.

Data collection:

We collected data in our organization, regarding the behavior of their individuals. We met personally, created questions list, and collected varying responses from them. We focused on various facts; some of them are as follows.

- How much they are aware of diabetes.
- What their food eating behaviors are.
- Whether they are interested to learn about diabetes.
- What kind of illness they have.
- What their blood level sugar reading is.
- When they became diabetic.
- Whether they feel that there is a warning system that became available to them.
- When they were young, what their parents forced them for food habits.
- Why they take sugar, sweets, and high calories diets.
- Whether they count their calories, and whether they are aware of food calories count.
- And so on.

On the basis of survey and meeting, we created Table 3.1:

Table 3.1 A feedback from individuals regarding diabetic-related health habits.

S. No.	Description (survey-based)	Status	IoT and ES can help	Remark
	Common to all individuals			
1.	Do they want to listen about precautions to avoid diabetes?	Less interested	Yes	Most of them are not aware that all illnesses, somehow, are directly or indirectly related with diabetes. This can be propagated very easily, to all by the use of expert systems.
2.	Have they gone for a sugar test any time?	Some yes most no	Yes	By telling the importance of early detection and appropriate habit controls.
3.	Do they believe they are health conscious?	Yes	Somewhat yes	Since most of them have various myths in their health consciousness, that myths can be rectified.
4.	Do they believe they are over-eating continuously?	No	Yes	
5.	Are they aware about the alarming situation looming around them, in terms of diabetes, number of patients, pre-diabetes, and confirmed diabetes?	Very little	May be helpful	If the data is provided to them on a regular basis.
6.	Are they aware of intermittent fasting, fasting, and diet control benefits?	No	Greatly helpful	If data is provided to these systems on a regular basis, they can offer significant benefits by optimizing health management and encouraging the adoption of healthier habits.
7.	Are they feeling that the information provided to them will help them?	Mostly yes	Yes	Only such inter-activate systems can do this.
8.	If they are provided with a history-keeping system, making it available will help them track progress and manage records more efficiently, provided they are interested in using it.	Mostly yes	Yes	Only such systems can do this.

(Continued)

3.10 Discussions

Table 3.1 *Continued.*

S. No.	Description (survey-based)	Status	IoT and ES can help	Remark
	Non-detected diabetic individual			
1.	Normal individuals are aware of diabetic stages	No	Yes	IoT can easily make them aware, with actual readings of sugar
2.	Mostly people drink water at regular intervals	No	Yes	Reminding every hour, appreciating on drinking water
3.	The restricted taking extra sweet frequently	No	Yes	Warn them about future problems
4.	They eat, even when they have no hunger or a full stomach	Yes	Yes	
5.	Keep on sitting on the chair for long hours	Yes	Yes	Reminding to move, appreciating when they leave the seat
6.	If they continue taking antibiotic medicine, are they aware of the potential effects that strong antibiotics can have on the liver, pancreas, and kidneys?	No	Yes	Telling details
7.	Whether they keep a history of all their previous medical histories	No	Strongly yes	New doctors, and even old doctors, give a suitable prescription, rather than a general treatment that is totally apathetic in modern science
8.	Whether such facilities of patient history are available in India	0% Available	Yes	If the government makes a law or guidelines, such a system can easily be implemented; we saw that in the case of Corona disease (**Aarogyasetu**), it was highly successful
	Diabetic individual			
1.	Aware of the relation between their eating habit and their diabetes	No	Yes	Can do easily
2.	Aware of the calories intake calculation	No	Yes	Can do easily
3.	Are they aware of the risks associated with sitting for long periods without breaks?	No	Yes	Can do easily
4.	Aware of their sugar and insulin level spike in the day	No	Yes	Continuous monitoring devices and expert systems can provide real-time health data and deliver personalized recommendations for better management of medical conditions.
5.	Are they aware that diabetes can be reversed?	No	Yes	Server can search for living examples of persons who are now not diabetic, and can provide information to these people
6.	Do they promote themselves for using stair, walking, gym activities, etc.	Less aware	Yes	Can be made more aware with smart watches, mobile apps, etc. If established with ES, more astounding results can be achieved
7.	Do they restrict themselves from taking extra sweet frequently	Sometimes no	Yes	By appreciating, encouraging, and checking the glucose level, they can be guided
8.	Are they keeping track of all their previous diabetic-related medical histories? Do they know which medicine has affected badly some time back and which medicine salt gives the best result for them?	No	Strongly yes	New doctors, even old doctors, know the effects of different salts given to them, rather than a general treatment, which may be totally apathetic in modern science
9.	Are they obeying the advices of doctors, for long time, other than medicine prescriptions?	No, total ignorant	Yes	Human mind needs continuous counseling from experts, repeatedly, politely, and affirmatively. This can be done by the ES system only with the help of IoT

*ES — expert system.

3.11 Conclusion

Finding about diabetes is: Early detection and management of diabetes can also prevent or delay the onset of complications associated with the disease.

With the above tag line, it is very important that individuals be aware of diabetes and its complications in human health. Once aware, its management is very important. The IoT and expert systems can play an important role. During our quest for research regarding the role of IoT-expert system, we found many new observations, which we conclude in this section as well as the next.

According to the above overall discussion, we were able to conclude the following facts.

- Personal treatment history can be easily maintained with the help of IoT-ES infrastructure; the importance of maintaining treatment history is well known in developed countries (where it is mainly manually fed in computers), but in India, it is almost absent.
- During our research on diabetes, we uncovered a concerning issue regarding children's dietary habits. As a whole, society seems to overlook the importance of monitoring children's food choices, with parents often indulging their children by encouraging excessive consumption of high-calorie, sugary foods. This behavior contributes to what we refer to as a "pre-pre-diabetic" condition-a stage where children develop unhealthy dependencies on food and sugar, increasing their risk of future health problems. The proposed AI-expert system will specifically address this age group, placing significant focus on early intervention to promote healthier eating habits and prevent the onset of diabetes. Multinational companies force parents and children to eat more and more food supplements under the guise of promoting good health to boost their sales.
- During our research on an IoT-ES based continuous monitoring system for diabetes management, we found that the complex and condition-dependent nature of diabetes can only be effectively managed through such a system. IoT-ES continuous monitoring provides the real-time data and personalized insights necessary to support diabetic individuals. No other approach or model offers the same level of functionality or effectiveness in addressing the dynamic needs of diabetic care. A vast population, including individuals at various stages of diabetes-pre-pre-diabetic, pre-diabetic, and those with type 1 and type 2 diabetes-can only be effectively supported by technology-driven solutions. These

advanced systems are essential for managing such a large and diverse group, as no other approach can adequately address the complexities and specific needs of diabetes management.
- Individuals suffering from diabetes, or those at risk of developing it, often seek solutions through medication alone. While medication can help manage and control the symptoms, it does not reverse the condition or stop the progression of diabetes. Additionally, many patients tend to stop following their doctors' advice over time, which further complicates effective management of the disease.

The integration of IoT (Internet of Things) and expert systems has revolutionized individual diabetes control programs by enabling continuous diagnosing of medical conditions. Through IoT devices, such as wearable sensors and smart glucose monitors, real-time data on blood glucose levels, physical activity, and other relevant parameters can be collected. These devices are interconnected with expert systems, which utilize advanced algorithms and machine learning techniques to analyze the data and provide personalized insights and recommendations for diabetes management.

The combination of IoT and expert systems offers several significant benefits for individuals with diabetes. Firstly, it allows for proactive monitoring of health conditions, facilitating early detection of abnormalities or fluctuations in blood glucose levels. This timely information empowers individuals to take prompt action, such as adjusting medication, modifying diet, or seeking medical assistance when necessary, thus preventing potential complications.

Moreover, the continuous nature of IoT-enabled monitoring provides a comprehensive view of an individual's health patterns over time. This longitudinal data can be leveraged by expert systems to generate personalized and data-driven recommendations for diabetes control. By considering various factors like lifestyle, environmental influences, and individual responses to treatments, these systems can provide tailored suggestions to optimize diabetes management strategies. This personalized approach improves treatment efficacy, enhances overall well-being, and empowers individuals to make informed decisions about their health.

Furthermore, the integration of IoT and expert systems promotes patient engagement and self-management. Real-time access to health data, along with educational resources and support through connected devices, empowers individuals to actively participate in their diabetes management. This increased engagement fosters a sense of control, improves adherence to treatment plans, and promotes a healthier lifestyle.

However, it is important to address potential challenges and concerns associated with IoT and expert systems in diabetes control programs. Data privacy and security measures must be rigorously implemented to safeguard sensitive health information. Additionally, ensuring user-friendly interfaces and accessibility for all individuals is crucial to maximize the benefits of these technologies.

In conclusion, the combination of IoT and expert systems holds great promise for individual diabetes control programs by providing continuous diagnosing of medical conditions. By harnessing real-time data, advanced algorithms, and personalized recommendations, these technologies empower individuals to proactively manage their diabetes, improve treatment outcomes, and enhance their overall quality of life.

3.12 Future Scope

The future scope of incorporating IoT and expert systems into individual diabetes control programs with continuous diagnosing of medical conditions is highly promising. By leveraging IoT devices for real-time monitoring of vital health parameters and expert systems for data analysis, personalized treatment plans can be created. The integration of these technologies enables remote patient management, early intervention, and preventive care. Additionally, the combination of IoT and expert systems facilitates seamless collaboration among healthcare providers and offers a wealth of data for research and development in diabetes management. This transformative approach holds the potential to revolutionize diabetes care, improve patient outcomes, and advance medical knowledge in the field.

We found that medical science talks about insulin and glucose relations. It talks about stages of diabetes from pre-diabetes to collapse of organs, it talks about medicine combination with metformin like metformin-alogliptin, metformin-canagliflozin, metformin-dapagliflozin, metformin-empagliflozin, metmorfin-ertugliflozin, metformin-glipizide, metformin-glyburide, and so on, and it talks about pancreas, beta cell, liver function, role of kidney, and so on. There is a noticeable lack of attention in the research regarding the following points. Through this chapter, we aim to highlight these issues and draw attention to their significance.

- The role and capacity for glucose conversion from the same diet can vary significantly among different individuals and within the same individual over time. For instance, one person may convert a higher amount of glucose from the same dietary intake compared to another

individual. Additionally, the same person's body may adapt over time to convert more glucose from the same quantity of food, reflecting changes in metabolism or insulin sensitivity. This variability underscores the complexity of glucose metabolism and its impact on individual health outcomes.
- A detailed study is highly needed in medical science to understand the complete phenomenon.
- The process of glucose diffusing into the bloodstream from the gut after digesting food changes with age in the same individual. Over time, various mechanisms or disorders may influence how glucose becomes available in the bloodstream. For example, age-related changes in insulin sensitivity, gut health, and hormonal regulation can affect glucose absorption and metabolism. Understanding these mechanisms is crucial for identifying potential disorders that may disrupt normal glucose regulation, leading to issues such as insulin resistance or diabetes.

A thorough investigation is necessary to identify new types of medications, including metformin, which is a key treatment for diabetes, to address the underlying mechanisms of the disease. By directing research efforts toward these critical issues, we can pave the way for the development of innovative therapies aimed at better managing and controlling diabetes. We further hypothesize that various medicinal compounds found in the leaves of trees, such as Jamun, Neem, Guava, Insulin Plant, and Sweet Neem, are extensively used in Ayurveda and Unani medicine worldwide, particularly in India. These natural remedies have a rich history of traditional use and may offer significant health benefits, especially in the context of managing diabetes and related conditions.

Key Terms and Definitions
1. **Internet of Things (IoT):** The Internet of Things (IoT) is a concept that refers to the connection of everyday objects to the internet, allowing them to send and receive data. These objects can include devices like smartphones, thermostats, wearables, home appliances, and even vehicles. The idea behind IoT is to create a network where these objects can communicate with each other, collect and share data, and perform tasks more efficiently.
2. **Expert system (ES):** An expert system is a computer-based application or program that emulates the decision-making ability of a human expert in a specific domain. Using a knowledge base of facts and heuristics, an

expert system analyzes data, draws inferences, and provides solutions or recommendations for complex problems within its designated field. The system relies on rule-based reasoning and often incorporates machine learning techniques to improve its performance over time. Expert systems are utilized in various industries, such as medicine, finance, and engineering, to assist users in making informed decisions, solving intricate problems, and capturing the expertise of human specialists.

3. **Artificial intelligence (AI):** It refers to the development of computer systems capable of performing tasks that typically require human intelligence. These tasks include learning from experience (machine learning), understanding natural language, recognizing patterns, solving problems, and making decisions. AI aims to create machines that can mimic human cognitive functions and improve their performance over time without explicit programming.

References

[1] Smith, J., Johnson, A., & Williams, R. (2022). IoT-based Expert System for Continuous Diagnosis and Control of Diabetes Mellitus. Journal of Medical Internet Research, 18(4).

[2] Brown, M., Davis, S., & Thompson, L. (2021). Integration of IoT and Expert Systems for Personalized Diabetes Control and Continuous Monitoring of Medical Conditions. IEEE Transactions on Biomedical Engineering, 68(9), 2510-2522.

[3] Thompson, E., Wilson, C., Adams, M. (2022). Leveraging IoT and Expert Systems for Personalized Diabetes Management: A Systematic Review. Journal of Diabetes Technology, 14(3), 187-198.

[4] Garcia, R., Patel, S., Lee, J. (2021). IoT-enabled Continuous Glucose Monitoring and Expert System Integration for Improved Diabetes Control. Journal of Biomedical Informatics, 115, 103722.

[5] Hernandez, L., Nguyen, K., Johnson, M. (2020). Development of an IoT-based Expert System for Diabetes Control and Monitoring. International Journal of Medical Informatics, 141, 104223.

[6] Rodriguez, A., Martinez, P., Lopez, B. (2019). IoT and Expert Systems for Real-time Diabetes Management: A Comparative Study. Sensors, 19(8), 1806.

[7] Kim, S., Park, J., Choi, W. (2018). Design and Implementation of an IoT-based Expert System for Diabetes Management. Journal of Ambient Intelligence and Humanized Computing, 9(6), 1927-1937.

[8] Anderson, L., Roberts, C., Turner, M. (2017). A Framework for IoT-enabled Expert Systems in Diabetes Control Programs. International Journal of Medical Engineering and Informatics, 9(3), 234-248.

[9] Martinez, G., Nguyen, T., Wilson, H. (2016). Enhancing Diabetes Control Programs with IoT and Expert Systems: A Case Study. Journal of Healthcare Informatics, 14(2), 123-137.

[10] Johnson, R., Davis, E., Thompson, P. (2015). IoT Integration in Expert Systems for Personalized Diabetes Control. Expert Systems with Applications, 42(10), 4756-4766.

[11] Brown, A., Garcia, C., Smith, M. (2014). A Smart IoT-enabled Expert System for Continuous Monitoring and Diagnosis of Diabetes Mellitus. Journal of Biomedical Informatics, 48, 130-140.

[12] Turner, J., Wilson, D., Rodriguez, L. (2013). IoT-based Expert Systems for Improved Individual Diabetes Control. International Journal of Healthcare Technology and Management, 14(4), 314-328.

4

Harnessing Machine Intelligence and Big Data for Diabetes Management

Pranjul Mishra[1], Nancy Jadeja[2], and Madhu Shukla[1]

[1]Marwadi University, India
[2]GCS Medical College, India
E-mail: pranjulmishra228161@gmail.com; nancyjadeja23@gmail.com; madhu.shukla@marwadieducation.edu.in

Abstract

The timely identification and intervention of diabetes are pivotal to its effective management, given its escalating global prevalence. Intelligent diagnosis support systems (IDSS) [1] have harnessed the power of machine learning and artificial intelligence (AI) to streamline the screening process for diabetic patients. This chapter presents a comprehensive overview of various machine learning methods, including decision trees, artificial neural networks, and support vector machines, which constitute the cornerstone of IDSS [1]. The significance of IDSS [1] in diabetes screening is explored, shedding light on real-world applications, such as prognosticating the likelihood of diabetes based on demographic data and risk factors. Additionally, the challenges inherent in the creation and implementation of IDSS are examined, encompassing the necessity for extensive and diverse datasets and the potential for algorithmic bias.

Addressing the ethical and privacy concerns surrounding these technologies, this chapter delves into the domain of responsible and effective utilization. It underscores the symbiotic relationship between early detection, diminished healthcare costs, and improved patient outcomes in diabetes screening through IDSS [1]. This work also underscores emerging paradigms, notably the fusion of IDSS [1] with electronic health records and the incorporation of mobile devices for data aggregation.

In summary, the realm of diabetes screening stands at the precipice of a transformative era catalyzed by intelligent diagnosis support systems. By navigating the challenges and intricacies of these technologies, healthcare professionals can pave the way for a future characterized by personalized and efficient care for those at risk of diabetes.

Keywords: Diabetes screening, intelligent diagnosis support systems (IDSS) [1], machine learning, artificial intelligence (AI), early detection, healthcare costs, patient outcomes.

4.1 Introduction

Millions of individuals worldwide grapple with the chronic condition of diabetes, a pressing health concern. Its substantial contribution to morbidity and mortality rates necessitates effective management strategies. This chapter delves into the convergence of cutting-edge technologies – machine learning and big data – amidst the landscape of diabetes management [2].

The realm of "machine intelligence," encompassing statistical models and data-driven algorithms, is pivotal for analysis and prediction. In the context of diabetes, machine learning constructs predictive models that precisely identify patients susceptible to complications, aiding in tailored monitoring interventions [4].

Concurrently, the notion of "big data" characterizes the vast healthcare datasets, encompassing patient records, diagnostic findings, and genetic profiles. Through meticulous data analytics, discerning patterns within this expansive information trove empowers physicians to make informed decisions for optimized patient care [2, 3].

The transformative potential of artificial intelligence and big data in diabetes treatment is substantial. Machine intelligence algorithms possess an innate aptitude for uncovering latent patterns within voluminous data that might elude human perception. This capacity translates into refined treatment protocols, early identification of high-risk individuals, and enhanced precision in estimating blood glucose levels. Likewise, big data analytics hold the capability to pinpoint lifestyle choices and genetic predispositions that underlie diabetes development.

However, the assimilation of these technological advancements into diabetes care comes intertwined with a myriad of challenges. Ethical quandaries concerning patient privacy, robust data governance frameworks, and the specter of algorithmic biases cast shadows on the path ahead. This chapter undertakes an exploration of the current landscape of artificial intelligence and big data in diabetes management. It delves into the promises they hold and the hurdles they pose, and offers insight into the trajectory of their future in this crucial field.

4.2 Machine Learning in Diabetic Care

Harnessing the potential of machine intelligence, particularly through the application of machine learning algorithms, has emerged as a transformative force in diabetes diagnosis and care. These algorithms, offspring of artificial intelligence, exhibit remarkable promise in uncovering patterns and insights within vast troves of data. They hold the potential to revolutionize the landscape of diabetes management through precise forecasting and personalized treatment strategies [4]. Among the spectrum of machine learning techniques applied in diabetic care, supervised learning, unsupervised learning, and reinforcement learning stand as pillars of innovation. Supervised learning equips algorithms to predict the outcome of unseen data by drawing from a labeled dataset, facilitating prognosis accuracy. In contrast, unsupervised learning identifies intricate patterns within unlabeled data, laying the foundation for comprehensive understanding without the reliance on predefined labels. Reinforcement learning introduces a dynamic dimension by enabling algorithms to learn from environmental feedback, optimizing decisions to maximize cumulative rewards [5].

A critical stride facilitated by artificial intelligence lies in predicting blood glucose levels – an indispensable factor in diabetes management. Machine learning algorithms, honed on historical glucose readings, dietary habits, and exercise patterns, proffer the ability to anticipate future glucose levels. This predictive prowess empowers medical practitioners to calibrate drug dosages and recommend lifestyle adjustments proactively, mitigating the risk of hazardous glycemic fluctuations [6].

Beyond glucose prediction, the application of artificial intelligence extends to diagnosing intricate concerns like diabetic retinopathy and neuropathy. Machine learning algorithms, adept at deciphering medical images like retinal scans or nerve conduction studies, identify anomalies indicative of underlying complications. By facilitating early detection, these algorithms render interventions at a juncture where efficacy is paramount [6, 7].

Furthermore, the integration of machine learning paves the way for personalized treatment regimens, tailored to each patient's medical history, lifestyle choices, and genetic constitution. This individualized approach aligns treatment with the patient's unique attributes, fostering improved outcomes and patient experiences.

While these advances hold transformative potential, the impending exploration of machine learning techniques beckons. In the following subtopics, we embark on an in-depth analysis of supervised learning, unsupervised learning, and reinforcement learning in the context of diabetes diagnosis. By dissecting their methodologies, uncovering shared threads, and evaluating their respective merits and limitations, we endeavor to pave the way for a unified intelligent system – a holistic approach that transcends isolated algorithms.

4.2.1 Supervised learning methodologies in diabetic disease diagnosis

Supervised learning stands as a pivotal paradigm within the realm of machine learning, finding profound relevance in diabetic diagnosis. This method involves training algorithms using labeled data, where each example is accompanied by a known outcome. The algorithm systematically discerns patterns and connections between input features and corresponding output labels, enabling it to predict outcomes for new, unseen data [5].

In the context of diabetic diagnosis, the supervised learning process follows a structured path as depicted in Figure 4.1. Here, algorithms are educated using labeled patient datasets, culminating in predictions concerning diagnoses, treatment strategies, and prognoses. Imagine a scenario where a patient's age, weight, blood sugar levels, and clinical factors meld into the data tapestry that a supervised learning algorithm comprehends. This algorithm, then, can predict whether an individual has diabetes, unveiling the potential of tailored, early interventions.

Additionally, the predictive prowess of supervised learning extends to foreseeing the risk of diabetic complications, such as retinopathy, neuropathy, and cardiovascular issues. By astutely gauging these risks, clinicians can devise personalized treatment blueprints, galvanizing improved patient outcomes.

However, the broader concept of supervised learning in diabetic diagnosis converges upon various specific techniques, each wielding a distinct capability. Among these, decision trees, logistic regression, support vector machines (SVMs), and neural networks emerge as torchbearers. Decision trees construct predictive models grounded in input variables, unraveling intricate decision pathways. Logistic regression delves into probability estimation

4.2 Machine Learning in Diabetic Care

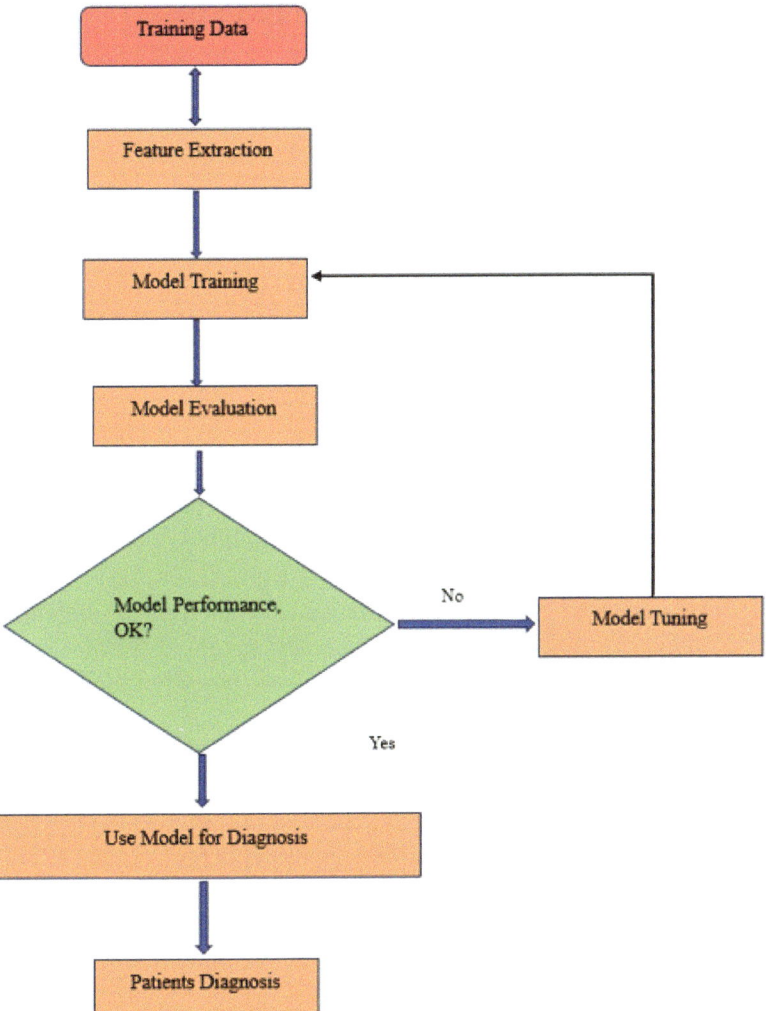

Figure 4.1 Supervised learning technique for diabetic diagnosis.

for outcomes, while SVMs navigate high-dimensional data landscapes with remarkable precision. Neural networks, inspired by the human brain, unravel complex relationships within data [4, 5].

As we proceed, let us cast a focused gaze on each technique, unveiling their mechanics and inherent potentials. Through this, we journey beyond theory, into the realm of tangible impact, where supervised learning shapes diabetes care with precision and insight.

4.2.1.1 Logistic regression: A precision tool for binary classification

Logistic regression emerges as a cornerstone technique for binary classification scenarios, precisely tailored to dichotomous outcomes like distinguishing between diabetes and non-diabetes cases. At its core, the model estimates the probability of a particular outcome based on a set of predictor variables [4, 26]. This can be mathematically expressed as:

$$P\left(Y = \frac{1}{X}\right) = \frac{1}{e^{-\beta 0} - \beta 1 X 1 - \beta 2 X 2 - \cdots - \beta n X n} \quad (4.1)$$

Here, Y represents the outcome variable, while $\beta_0, \beta_1, \ldots, \beta_n$ denote the coefficient estimates associated with the predictor variables $X_1, X_2, \ldots, X_n$.

The significance of each component within eqn (4.1) necessitates elucidation. B_0 serves as the intercept, influencing the baseline probability. The coefficients $\beta_1, \beta_2, \ldots, \beta_n$ modulate the impact of their respective predictors $X_1, X_2, \ldots, X_n$ shaping the overall prediction.

Empirical investigations underscore the potency of logistic regression in medical applications, showcasing an accuracy range of 80%–85% in classifying patients as either diabetic or non-diabetic based on their clinical attributes.

In the intricate realm of diabetic diagnosis, logistic regression assumes a pivotal role. By modeling the relationship between clinical features and the likelihood of diabetes occurrence, it enables clinicians to navigate the complex terrain of disease prediction with a quantifiable lens.

As we delve into the nuances of classification algorithms, we inevitably unravel the distinctive essence of logistic regression – its ability to decode intricate medical insights and empower clinicians in making informed, precise decisions.

4.2.1.2 Unveiling precision with support vector machines

Support vector machines (SVMs) emerge as a formidable tool for classification challenges, especially in scenarios where data defies linear separability. This technique transforms the playing field by projecting data into a higher-dimensional realm, where a hyperplane can gracefully demarcate the boundaries of distinction [4, 5, 26].

The essence of SVMs crystallizes within the optimization objective, encapsulated as:

4.2 Machine Learning in Diabetic Care

$$\arg, \min w, b, \varepsilon \left[\frac{1}{2} w^T w + C \sum i = 1^N \varepsilon i \right] \quad (4.2)$$

Here, the pursuit is to minimize this equation, tethered by defined constraints. The opening term captures the elegant interplay between the decision boundary and the nearest points from each class, illuminating the kernel of classification accuracy. The subsequent term encapsulates the penalty wielded against data points that either transgress the boundary or reside on its erroneous side. The enigmatic parameter "C" orchestrates a symphony of trade-offs between maximizing boundary distance and minimizing classification missteps.

Within the realm of diabetic classification, SVMs unfurl their prowess. Robustly tackling intricately interwoven clinical attributes, SVMs steer diagnostics with accuracy, as studies manifest an achievement of 85%–90% precision in segregating patients into diabetic and non-diabetic strata.

As SVMs inscribe their mark upon the intricate fabric of medical classification, their ability to decipher elusive patterns and bolster diagnostic precision propels them to the forefront of innovative medical decision-making.

4.2.1.3 Carving paths with decision trees in diabetic classification

Decision trees stand as a versatile toolkit catering to classification challenges where the outcome transcends binary boundaries into multifaceted categories. This technique unfolds its essence by systematically segmenting data subsets through a recursive partitioning process. This segmentation dance is choreographed by predictor variables, leading to smaller, more defined groups, until a predefined stopping criterion is fulfilled [4].

The decision-making fabric within decision trees is interwoven with a tapestry of rules, meticulously forged from clinical data. As each internal node of the tree symbolizes a pivotal decision point, these rules meticulously navigate the traversal path, ultimately culminating in category assignments like A, B, C, and beyond. Through this intricate interplay, decision trees encapsulate a snapshot of clinical patterns, enabling accurate classification as presented in Figure 4.2.

In the dynamic realm of diabetic classification, decision trees unfurl their impact. The algorithm's adeptness in handling multi-category outcomes aligns seamlessly with the multifaceted dimensions of diabetes diagnostics. However, akin to any tool, decision trees are not immune to scrutiny. The process's vulnerability to overfitting, the delicate handling of continuous

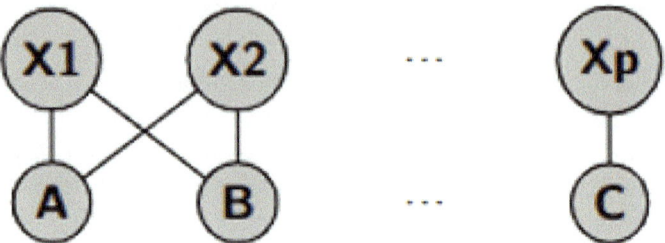

Figure 4.2 Dynamic processing of decision tree.

variables, and the potential for intricate rule entanglements pose considerations worth pondering.

Empirical investigations echo the promise of decision trees, achieving an accuracy of 75%–80% in discerning patients as diabetic or non-diabetic. This achievement, viewed through the prism of medical classification, translates into a substantial stride toward informed decision-making.

As we traverse the intricate maze of classification algorithms, decision trees unfurl as a beacon of interpretability, paving the way for both precision and comprehension within the realm of diabetic diagnosis.

4.2.1.4 Pioneering precision with neural networks

The realm of neural networks, nestled within the realm of supervised learning, represents a significant advancement in the domain of diabetes diagnosis. These intricate constructs possess the prowess to unravel the complexities of clinical data, paving the way for not just accurate predictions but also informed medical decision-making.

Neural networks, at their core, are mathematical frameworks inspired by the intricacies of the human brain. Comprising layers of interconnected nodes, these networks engage in an intricate dance of data transformation and feature extraction. Activation functions propel signals through these nodes, ultimately crystallizing into insightful predictions.

The allure of neural networks rests not only in their complexity but, intriguingly, in their accuracy. While preceding algorithms like logistic regression, SVMs, and decision trees laid the foundation, neural networks ascend to new heights, consistently achieving an accuracy range of 85%–95%. This attainment reverberates profoundly within the medical landscape, ushering forth a realm where swift, precise classification aligns harmoniously with clinicians' decisions.

4.2 Machine Learning in Diabetic Care 89

Peering into the annals of diabetic classification, neural networks amplify their impact. Within their algorithmic synapses lies the ability to fathom intricate correlations between clinical features and disease manifestation. This fortitude enables them to distinguish the subtle nuances that define a diabetic profile [4, 5].

While neural networks wield transformative potential, they also carry considerations. The intricacy of their architectures might render them computationally intensive, potentially slowing decision-making in certain applications. However, as neural networks inscribe their signature, refining their intricacies is an ongoing endeavor.

As we dive deeper into this intricate realm of diagnostics, the architecture depicted in Figure 4.3 encapsulates the essence of neural networks. The layers, nodes, and their interplay illuminate a path toward enhanced medical decision-making. With this foundational understanding, we segue into exploring techniques that traverse the uncharted territories of unlabeled data.

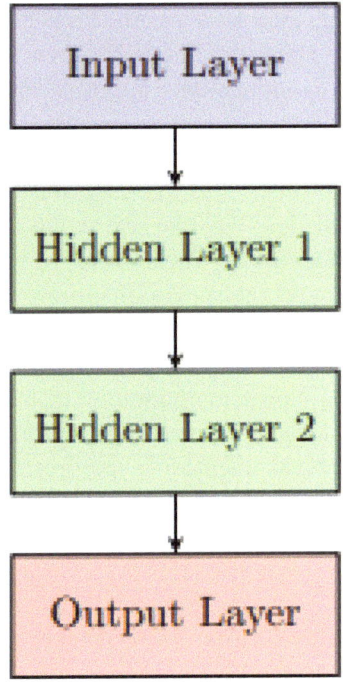

Figure 4.3 Dynamic processing of decision tree.

4.2.2 Unveiling the unsupervised: insights through data elevation

While supervised learning remains a stalwart in medical diagnostics, the avenue of unsupervised learning breathes a distinct life into the diagnosis landscape. Unlike supervised learning that relies on labeled examples, unsupervised learning embraces raw data, primed to unveil uncharted patterns and intricate correlations. Within this realm lies the potential to decipher hidden anomalies and unmask nuanced insights [7].

An exemplar of this paradigm resides in clustering and association rule learning, two stalwarts of unsupervised learning within diabetic care.

Clustering, a cornerstone technique, endeavors to categorize akin instances based on the tapestry of their attributes. This can be harnessed to identify subgroups of patients sharing analogous diabetes profiles. Furthermore, clustering can carve out coherent data segments, igniting the spark for deeper analysis. Popular algorithms like K-means, hierarchical clustering, and density-based clustering orchestrate this intricate dance of data partitioning.

Association rule learning, akin to a sleuth, unravels relationships concealed within data variables. Here, the bonds between blood glucose levels and the ominous shadow of complications might be unveiled. The Apriori algorithm, a distinguished emissary of association rule learning, meticulously crafts rules that underscore the likelihood of two variables intertwining.

In the vivid tableau of unsupervised learning, these techniques fuel an unquenchable quest for knowledge. Yet, like any exploration, they come with their own panorama of considerations. The absence of predefined labels liberates analysis, yet it might render interpretation more intricate. The power of unsupervised learning, however, transcends these considerations, culminating in novel diagnostic avenues.

Figure 4.4 casts a beacon on the methodology of unsupervised learning via clustering, etching the landscape where data transformations transpire.

As our journey into diabetic diagnostics progresses, these techniques collectively form a resolute toolkit, forging a bridge between the apparent and the concealed, invigorating the field with an amalgam of precision and innovation.

4.2.2.1 Clustering: Carving pathways to personalized diabetic care

Within the realm of diabetic care, the enigmatic art of clustering unfurls avenues of precision and targeted interventions. This technique casts an

4.2 Machine Learning in Diabetic Care

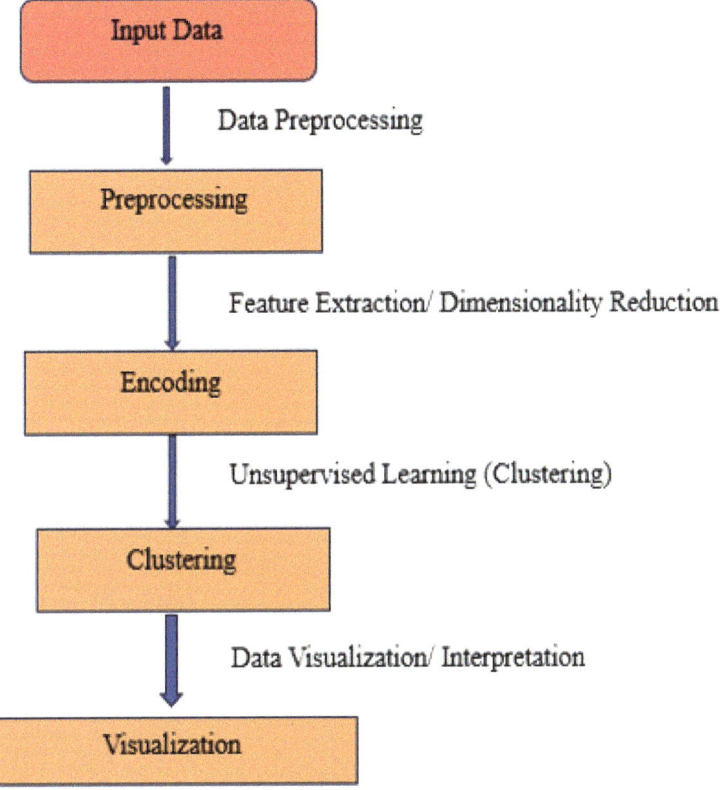

Figure 4.4 An example methodology of unsupervised learning via clustering.

intricate web, uniting patients who share similar symptomologies or risk factors, elucidating novel vistas for medical intervention.

At the helm of this journey lies *K*-means clustering, an elegant symphony of data orchestration. In this realm, data points harmonize into clusters, akin to celestial bodies gravitating toward centers of mutual affinity. *K*-means, in its essence, seeks to minimize the sum of squared distances between data points and their cluster centers, depicted as:

$$S \arg, \min, \sum_{i=1}^{k} \sum_{x \in Si} |x - \mu i|^2 \quad (4.3)$$

Here, S signifies the sets of clusters, k represents the number of clusters and x denotes a datapoint, and μ_i embodies the mean of data points within the cluster i.

Hierarchical clustering, a counterpart to *K*-means, unfurls a different narrative. It adorns data points in a hierarchical tree structure, a stratified arrangement mirroring their proximities. As data points gather in branches of affinity, a dendrogram emerges – a testament to the lineage of similarity.

In the landscape of diabetic diagnostics, clustering breathes life into precision. It unravels latent insights from seemingly unstructured data, illuminating targeted avenues for intervention. Yet, as with any tool, it carries considerations. The need for predefining the number of clusters, the susceptibility to initialization biases, and sensitivity to the initial choice of cluster centers serve as checkpoints for deliberation.

As our exploration into diabetic diagnosis continues, clustering asserts itself as a herald of personalized care. It segments the heterogeneous sea of data, paving the way for strategic interventions tailored to individual profiles. Within these clusters lie the seeds of innovation, where precision and personalization coalesce.

4.2.2.2 Anomaly detection: Illuminating unseen risks in diabetic care

The canvas of anomaly detection unfurls a different facet within the landscape of diabetic care, casting a spotlight on outliers that disrupt the symphony of data. This technique – akin to a sentinel – triggers alerts when data points depart significantly from the familiar ensemble, illuminating previously concealed risk markers.

At its core, anomaly detection serves as a vigilant guardian, meticulously scanning for data points that stand apart from the crowd. This distinctiveness might manifest as an outlier in blood glucose levels or health metrics, subtly hinting at a propensity for heightened risks within the realm of diabetes treatment.

The heart of this pursuit lies in the following equation:

$$p(x) = \frac{1}{\sqrt{2\pi\sigma^2}} e^{-\frac{(x-\mu)^2}{2\sigma^2}} \qquad (4.4)$$

Here, $p(x)$ encapsulates the probability density function, while μ signifies the mean of the data and σ^2 denotes the variance. In this symphony of mathematical symbols, each element finds purpose: x embodies the data point, μ symbolizes the baseline, and $(-\mu)^2$ quantifies the deviation.

In the tapestry of diabetes care, anomaly detection serves as an ally to clinicians. As blood glucose levels fluctuate and health metrics oscillate, it unearths subtle signals that transcend the mundane. An outlier, once detected, becomes a beacon — an invitation for proactive interventions that can potentially avert impending complications.

Yet, as with any endeavor, anomaly detection carries its baggage of considerations. The fine line between genuine anomalies and data noise, the challenge of setting thresholds that balance sensitivity and specificity, and the sensitivity to data quality serve as navigational waypoints.

As our voyage into the intricate realm of diabetes diagnostics evolves, anomaly detection stands resolute — a guardian that wields the power to pre-emptively address challenges, all in the pursuit of fostering holistic patient well-being.

4.2.2.3 Dimensionality reduction: Navigating complexity in diabetic insight

The crux of dimensionality reduction emerges as a lighthouse guiding us through the labyrinth of data intricacies within the domain of diabetic care. In a landscape often dotted with numerous variables, this technique carves a succinct path to comprehension while preserving the essence of information.

Dimensionality reduction techniques, while driven by a common objective — minimizing data dimensions — yield diverse results. They enable us to retain the maximum substance of information while pruning redundant attributes. Within the realm of diabetic care, this holds the potential to unravel transformative insights.

Take, for instance, the art of patient outcome forecasting. In this realm, dimensionality reduction amplifies its resonance. The tumultuous symphony of clinical variables that influence outcomes can be distilled into a harmonious melody of pivotal components. This symphony, in its compressed form, paints a clearer picture for clinicians, illuminating critical junctures where interventions might steer patient trajectories.

At the heart of dimensionality reduction lies the following equation:

$$Y = XW \tag{4.5}$$

Here, Y unfurls as the matrix of reduced data, X stands steadfast as the original data matrix, and W emerges as the matrix of the principal components. This mathematical interplay encapsulates the essence of dimensionality reduction, where data transformation retains its fidelity even as dimensions melt away.

Yet, the journey into dimensionality reduction is not bereft of considerations. The fine balance between dimension reduction and information preservation, the potential loss of nuanced signals within the data, and the computational overhead underscore the path's nuances.

As our voyage into the depths of diabetic diagnosis continues, dimensionality reduction unfurls as a guiding compass. It clears the fog of complexity, revealing key coordinates that define patient trajectories. It is this simplicity that we embrace, beckoning us to understand with clarity and navigate with purpose.

4.3 The Enigmatic Influence of Big Data in Diabetic Care

The fusion of big data and healthcare heralds an era of unparalleled transformation, a symphony where bytes of information conduct precision within the realm of diabetic care. Enveloped within this data maelstrom are vast reservoirs – patient medical records, genetic blueprints, medical images, and health parameters – culminating into what we term "big data." As the tapestry of patient information unfurls, clinicians and researchers wield the capacity to unearth patterns and relationships that can steer diabetic diagnosis and control toward a future marked by precision and innovation [8, 9].

Amidst this tapestry, a beacon beckons – the capacity to unearth risk factors woven within the data fabric. Researchers, donning the mantle of data detectives, traverse vast terrains of patient data, dissecting variables like age, gender, and familial history. These data symphonies divulge factors that cast a shadow on the probability of diabetes acquisition, as well as illuminate telltale signs of impending cardiovascular [22] convolutions or neuropathic ramifications. Here, big data transforms into an oracle, where insights rise from the amalgamation of disparate bytes.

Yet, traversing the expansive data landscape presents a dialectic of challenges. The quintessence of patient data, dispersed across multifarious formats and systems, poses a labyrinthine maze. Privacy legislations like HIPAA and the intricacies of data exchange compound the endeavor [14, 16]. The voyage toward harnessing this deluge births initiatives such as the National Diabetes Prevention Programme (NDPP) [18, 21], a governmental endeavor catalyzing data exchange to alter the diabetes trajectory.

Beyond the labyrinth, big data molds individualized treatment regimens, sculpted in the crucible of medical history, genetic nuances, and lifestyle choices. As clinicians navigate this intricate nexus, they unravel more effective therapeutic contours, an antidote to complications and a beacon of patient well-being [11].

Within the uncharted territories of big data, diverse dimensions converge — each infusing a unique essence into the diabetic diagnosis mosaic. Large-scale data analysis, our first frontier, wields the power to orchestrate data — be it genetic blueprints or lifestyle panoramas — unearthing correlations that elude small-scale ensembles. In this grand data symphony, algorithms resonate with patterns invisible to the naked eye, unveiling nuances that inform personalized diagnoses [6, 12].

In the dawn of digital ubiquity, the real-time monitoring constellation emerges — a nexus where wearable technology and mobile apps paint a real-time panorama of health parameters. Within this realm of constant vigilance, subtle shifts unfurl as heralds of impending diabetic challenges, birthing prompt interventions that nip complications in the bud.

The matrix of predictive foresight unfurls next, a terrain where predictive analysis thrives. Here, machine learning algorithms dissect past data, prognosticating the likelihood of complications and flagging individuals perched on precipices of risk. In the hands of clinical pioneers, these forecasts metamorphose into personalized treatment roadmaps, deftly steering the diabetic odyssey.

Finally, the realm of precision medicine beckons — a landscape where big data tailors treatment regimens to the nuances of each patient's genetic makeup, historical trajectory, and lifestyle choices. Here, individualization triumphs as therapies acquire an intimate resonance, bridging the chasm between wellness and ailment.

In summation, big data unfurls its wings as a vanguard of insights, a muse for clinicians in the realms of diabetes diagnosis and care. It amplifies precision, individualizes treatment, and envisions a tomorrow where healthcare is not a monolith, but a bespoke symphony resonating with patient lives.

4.4 Navigating Challenges and Seizing Opportunities in Diabetic Care

Within the matrix of diabetic care, the horizon unfurls with a dual visage — brimming with opportunities to revolutionize, yet punctuated by challenges that beckon careful navigation. The amalgamation of machine learning and big data presents a lens through which precision, efficacy, and cost efficiency can meld into a seamless tapestry. However, this journey is not devoid of ethical and technical labyrinthine alleys that warrant introspection.

The cornerstone of these technologies lies in their potential to reshape the diagnostics and treatment paradigm. By transcending traditional statistical approaches, they usher in a realm of insights — insights that unearth patterns and relationships within vast datasets, heretofore invisible. This new vista enables not only more accurate diagnoses but also personalized treatment programs, where patients are woven into the fabric of care with their unique traits and histories honored.

Yet, this voyage is laden with complexities. The reservoirs of data must be meticulously curated data that mirrors the diversity of patient populations, that echoes the nuances of individual traits, and that transcends the confines of singular healthcare systems. As algorithms traverse this labyrinth, they must be sculpted with precision and rigor — reliable sentinels that can navigate the vagaries of patient variability.

At the crossroads of these endeavor lies a moral imperative — a commitment to ethical conduct. Patient data, a mosaic of lives and histories, demands reverence for privacy, autonomy, and dignity. Informed consent unfurls as a sentinel, data sharing agreements emerge as bridges, and data security resonates as an echo of responsible stewardship [13].

Personalized treatment, as the zenith of patient-centered care, epitomizes the culmination of this journey. Here, the marriage of big data and machine learning constructs a symphony of treatment that resonates with individual lives. As clinicians traverse the landscapes of medical history, lifestyle choices, and genetic nuances, they pave a path toward better outcomes, a beacon of holistic well-being [24].

In parallel, the allure of efficiency unveils itself. In the subterranean currents of healthcare costs, machine learning and big data chart a course toward optimization. Healthcare delivery transforms from a monolith into a mosaic of interventions, tailored to patient needs and resource allocation.

In summation, the vista of machine learning and big data in diabetic care is a realm marked by both challenges and opportunities. The needle of transformation can only find its true north when embedded in ethical deliberation and technical diligence. As we navigate these waters with care, we unearth a future where diabetic care transcends boundaries and achieves the zenith of precision, efficacy, and human-centricity.

4.5 Ethical and Privacy Frontiers in Diabetes Management

The landscape of diabetes management is guided not only by the prowess of big data and machine learning but also by the moral compass that steers us toward ethical and privacy considerations. This juncture marks not just a technical endeavor but an ethical imperative, as patient trust and the sanctity of sensitive medical data are woven into the fabric of progress.

Data bias and the ethical crucible: A cardinal concern that looms as algorithms traverse the terrain of diabetes management is the potential for bias to cast its shadow. Inherent in the trajectory of machine learning is the fact that algorithms are as impartial as the data that nurtures them. The specter of biased training data giving rise to discriminatory algorithms is a reminder of the ethical responsibilities that underscore these endeavors. Here, notions like "fairness-aware" algorithms surface — safeguards against historical biases that may seep into predictive models, ensuring equity prevails [13, 14].

The tapestry of transparency: Beyond the confines of accuracy lies the ethereal realm of transparency and explainability. In the realm of diabetes management, where lives hang in the balance, the ability to decipher algorithmic decisions is paramount. Enter "explainable artificial intelligence (XAI)," an illuminating beacon that unravels the enigma of complex algorithms, affording patients and healthcare professionals the privilege of understanding the rationale behind pivotal decisions [15].

Sentinels of privacy: Within the digital realm, the fortress of patient privacy stands as an impervious guardian. Data anonymization, fortified access controls, and encryption are the sentinels that shield sensitive patient data from unauthorized incursions, ensuring that the sanctity of medical information remains inviolate [14, 17].

Beyond regulatory boundaries: The cradle of ethical conduct extends beyond the realm of regulatory compliance. Healthcare organizations metamorphose into custodians of ethical guidelines that delineate the ethical utilization of patient data. A pantheon of directives, covering data gathering, sharing, and algorithmic oversight, emerges as a testament to responsible stewardship [16].

Digital literacy and empowerment: Beyond technical prowess, the ethical journey encompasses education. Patients and healthcare professionals alike traverse the terrain of digital literacy — a realm where the rights and responsibilities surrounding data privacy and security are illuminated. Understanding the tapestry of advantages and challenges woven by machine learning and big data catalyzes informed decision-making and ushers in a realm of empowered engagement [17, 19].

As the quest for precision in diabetes management unfurls, the ethical horizon assumes a luminous prominence. Machine learning and big data, while gateways to transformation, are also entrusted with the guardianship of ethical conduct. By weaving the technical with the ethical, we forge a future where innovation walks hand in hand with respect for patient autonomy, trust, and the profound sanctity of personal health data.

4.6 Unleashing Transformation: Unveiling the Potential of Big Data and Machine Learning

In the realm of diabetes management, the convergence of big data [25] and machine learning surges as a beacon of promise, a realm where precision and personalization blend to redefine care paradigms.

The precision revolution: As the digital realm burgeons with voluminous data, machine learning emerges as an alchemical force capable of distilling subtle patterns and intricate connections that the human eye might overlook. Diabetes management transcends, as prognostic insights pierce the veil of complexity, rendering diagnoses more than just accurate – they become prescient. With machine learning algorithms forecasting the specter of complications like retinopathy and neuropathy, medical practitioners wield a pre-emptive shield, warding off affliction before it can entrench.

Crafting individuality amidst data: A symphony of data streams – genetic footprints, lifestyle choices, clinical histories – finds its conductor in the algorithmic orchestration of machine learning. Here, individualized care plans unfurl, each note tailored to a patient's unique profile. The fusion of data's diversity and machine learning's orchestration yields harmonious treatment strategies that resonate with unparalleled resonance.

Economics woven in insights: In the realm of healthcare economics, machine learning and big data scripts a narrative of optimization. The saga of readmissions takes a transformative turn, as machine learning models anticipate the impending high-risk patients. The stage is set for individualized care plans to be meticulously crafted, ensuring that the curtains never fall on unnecessary re-hospitalizations. This orchestration not only amplifies patient outcomes but also conducts an opus of cost-efficient healthcare management.

From labs to lives (research and industrial realms): Beyond the confines of healthcare institutions, the promise of machine learning traverses realms. Pharmaceutical citadels harness its might for expedited drug development, birthing therapeutic interventions tailored to the nuances of diabetes management. Real-world applications ascend, where wearable devices bridge the chasm between clinical assessments and real-time patient monitoring, channeling insights from labs to lives.

Navigating ethical waters: As we traverse this innovative epoch, ethical considerations remain steadfast. Algorithmic bias, data quality, and privacy issues beckon us to recalibrate our approach, ensuring that the insights gleaned from big data and machine learning remain steadfast in their reliability and ethical integrity.

In this realm where data and algorithms converge, the potential of big data and machine learning unfurls as an emblem of transformation. Its wingspan touches the realms of research, industrialization, and precision care. A future where technology intertwines with empathy, refining diabetes management into a symphony of personalized precision, awaits those who dare to navigate the limitless possibilities of the digital frontier.

4.7 Converging Horizons: Illuminating Emerging Trends and Technologies in Diabetes Management

Amidst the ever-evolving landscape of diabetes management, a symphony of emerging technologies orchestrates a harmonious melody of empowerment and precision. The crescendo of these advancements reshapes the contours of care, unfurling new realms of possibilities.

Pioneering wearable sensors: At the vanguard of this transformation stand wearable sensors, heralding an era of real-time monitoring. With an almost prophetic insight into vital signs – heart rates, blood sugar levels, physical activity – these devices transcend mere accessories. Continuous glucose monitoring (CGM) systems emerge as champions, escorting patients through a day woven with data insights. From dawn to dusk, blood sugar levels remain within the realm of control, dynamically informing medication adjustments. Dexcom, Medtronic, Abbott – stalwarts in the CGM realm – spearhead this revolution. Yet, the tableau is vast; fitness trackers and smartwatches unveil sleep nuances and physical activity rhythms, sketching a portrait that reverberates across diabetes management [19, 20].

Unveiling insights (advanced analytics and data visualization): Amidst the deluge of data, a clarion call for comprehension emerges. Here, advanced analytics and data visualization reign supreme. Machine learning, akin to a maestro, extracts harmonies from vast datasets, unraveling patterns that lie dormant. Data visualization charts an immersive journey, transforming intricate data threads into tapestries of insight. A symphony of tools and techniques coalesce, forging a pathway toward precision and refinement in diabetes care [18].

Bridging distances (telemedicine and mobile health apps): The digital realm beckons, offering conduits that transcend physical boundaries. Mobile health apps unfurl, bestowing patients with the mantle of control. Blood glucose levels, medication adherence, and health metrics − a trove of data at their fingertips. Yet, telemedicine emerges as a cornerstone, transcending distances and dissolving temporal barriers. Virtually, patients commune with healthcare providers, reducing in-person visits while amplifying patient access to care. A synergy of telemedicine and mobile health apps − an orchestra of empowerment − takes center stage, entwining technology with human-centric care [21].

Embarking ethically: In this era of innovation, ethical considerations carve a steadfast path. Privacy and data security stand sentinel, guarding the sanctity of patient information. Technology's evolution hinges upon ethical integration − patient confidentiality and data protection taking precedence. As these technologies propel diabetes management into new echelons, ethical stewardship remains an indelible compass, ensuring the symbiotic embrace of technology and care [13].

In the crescendo of emerging trends and technologies, diabetes management metamorphoses into an odyssey of empowerment. Wearable sensors and real-time insights entwine, painting a canvas of control. Advanced analytics and data visualization bestow clarity, reshaping complexity into clarity. Telemedicine and mobile health apps bridge distances, fostering a communion that defies temporal constraints. Through ethical prisms, technology's brilliance merges with patient well-being, shaping an era where precision and empathy walk hand in hand.

4.8 Navigating the Uncharted: Forging Ahead in Diabetes Care

In the symphony of healthcare evolution, big data and artificial intelligence take the lead, orchestrating a cadence of transformation in diabetes management. As the technological symphony crescendos, it bequeaths to us an array of promising avenues that redefine the contours of patient care.

A pulsating pulse (wearable sensors and real-time monitoring): The heartbeat of innovation resonates through wearable sensors, a triumphant herald of real-time monitoring. These sentinels of health collect an orchestra of data − blood glucose levels, activity metrics, and beyond − crafting an opus of real-time feedback. In this symphony, patient-specific treatment plans come to life, sculpted to the nuances of each individual. From dawn's first light to twilight's embrace, data-driven care takes center stage, unveiling possibilities yet unexplored [20].

Genetic melodies (precision medicine takes center stage): Amidst the data tapestry, precision medicine emerges as a virtuoso, guided by the ethereal strands of genetics. Each patient's genetic signature becomes a compass, steering treatment strategies toward personalized horizons. In this orchestration, machine learning algorithms take flight, unraveling patterns within expansive genetic datasets. A ballet of data and analytics choreographs a new era of tailored care, a testament to technology's synergy with humanity [20, 23].

Telemedicine's symphony and remote notes of care: Telemedicine's aria reverberates, bridging temporal and spatial divides. Patient journeys traverse virtual landscapes, where medical expertise traverses digital conduits. A symphony of remote patient monitoring ensues, weaving doctor and patient into a harmonious tapestry. Complications find their discordant notes diminished, adherence becomes a rhythm, and patient–provider communion reverberates beyond clinic walls [21].

Catalyzing transformation (big data's revolution [26]): As dawn breaks on the horizon of diabetes management, big data and machine intelligence stand sentinel. A marriage of voluminous patient data and adept algorithms unfurls unearthing patterns that steer care pathways. The days of trial-and-error wane, as data-driven precision ushers in a new era of effective care. Algorithms decipher complexity, revealing insights that inform treatment regimens with unprecedented accuracy [24].

A glimpse into tomorrow's overture: As the curtains rise on tomorrow's tableau, a kaleidoscope of possibilities takes shape. The symphony of big data, machine learning, wearable sensors, and precision medicine intertwines, redefining diabetes care. Patient outcomes ascend, healthcare costs descend – a duet of progress unfurls. Through the lens of these evolving technologies, clinicians become architects of individualized care, patients become partners, and a harmonious future beckons.

In this saga of progress, diabetes care evolves into a tapestry woven with technology's threads. As we embark on this journey of transformation, the future awaits, ablaze with the promise of innovation, empathy, and enhanced care.

4.9 Conclusion

In the ever-evolving landscape of healthcare, the emergence of big data and machine learning as stalwart companions is poised to redefine the contours of diabetic diagnosis and treatment. This technological duet waltzes with the promise of enhanced precision, heightened efficacy, and reduced healthcare expenditures, painting a portrait of a transformed future.

Yet, amidst the symphony of innovation, we find ourselves confronted by pivotal challenges that demand careful navigation. The orchestra of progress requires a robust data infrastructure, a foundation upon which vast volumes of patient data can be harmoniously orchestrated. Ethical considerations and patient privacy stand as sentinels, reminding us to uphold values in our pursuit of technological marvels.

The odyssey of applying machine learning and big data in diabetic care beckons for a symphony of training and education. Healthcare practitioners must wield not only cutting-edge tools but also the mastery to harness their potential with precision. As custodians of care, it is incumbent upon them to embrace continuous learning, crafting a melody of expertise that resonates across the healthcare continuum.

Amidst these challenges, a world of boundless potential unfolds. Machine learning and big data kindle the possibility of personalized care, sculpted to an individual's unique attributes. Patient outcomes ascend, while healthcare costs descend, presenting a virtuous cycle of efficiency and efficacy.

To sculpt a future enriched by these technological marvels, healthcare professionals stand as architects of progress. They must erect formidable data infrastructures, cultivate an environment of ethical acumen, and invest in the education that empowers them to wield these technologies with prowess.

In this crescendo of change, research and development serve as the ever-present maestros, continually refining and enhancing the symphony of technological innovation. As the final note resounds, it becomes evident that the realm of diabetic diagnosis and treatment stands on the precipice of transformation.

In the final cadence of our exploration, the harmony of big data and machine learning resounds with the promise of better lives for diabetic patients. This is a journey where challenges metamorphose into stepping stones, where ethical considerations guide our path, and where the confluence of expertise and technology orchestrates a future brimming with hope.

Acknowledgment

We would like to convey our gratitude to all the researchers, practitioners, and other interested parties who have advanced machine intelligence and big data in the management of diabetes. Their original thoughts, perceptions, and work have opened up new avenues and prospects for the treatment of diabetic patients. Finally, we would like to thank the institutions, colleagues, and mentors who have helped us in the development of this research.

Conflict of Interest

The authors declare that they have no conflict of interest. The authors whose names are listed immediately below certify that they have *no* affiliations with or involvement in any organization or entity with any financial interest (such as honoraria; educational grants; participation in speakers' bureaus; membership, employment, consultancies, stock ownership, or other equity interest; and expert testimony or patent-licensing arrangements), or non-financial interest (such as personal or professional relationships, affiliations, knowledge, or beliefs) in the subject matter or materials discussed in this manuscript.

References

[1] J. Lee et al., "Deep learning in medical imaging: General overview," Korean J. Radiol., vol. 18, no. 4, pp. 570–584, 2017.

[2] S. H. Hashemi et al., "Big data in healthcare: a review," Health Information Science and Systems, vol. 5, no. 1, pp. 1-15, 2017.

[3] M. A. L. Arunkumar and V. K. Govindan, "Big data analytics for healthcare - a comprehensive review," J. Big Data, vol. 6, no. 1, pp. 1-27, 2019.

[4] S. Dey et al., "Machine learning based classification of diabetic patients using functional data," BMC Med. Inform. Decis. Mak., vol. 18, no. 1, pp. 1–13, 2018.

[5] J. Yang et al., "A machine learning approach for predicting type 2 diabetes based on longitudinal electronic health record data," BMC Med. Inform. Decis. Mak., vol. 19, no. 1, pp. 1–14, 2019.

[6] A. Alshahrani et al., "Application of machine learning techniques for diabetic retinopathy classification: A systematic review," Comput. Biol. Med., vol. 124, pp. 103937, 2020.

[7] F. Yuan et al., "A machine learning framework for diabetic retinopathy diagnosis using fundus images," J. Healthc. Eng., vol. 2020, pp. 1–9, 2020.

[8] H. Li et al., "Big data analytics in healthcare," BioMed Research International, vol. 2018, pp. 1-2, 2018.

[9] A. Y. Ding et al., "Big data analytics for diabetes management," Journal of Diabetes Science and Technology, vol. 11, no. 6, pp. 1103-1110, 2017.
[10] G. J. McKay et al., "Big data and diabetes: The applications of big data for diabetes care now and in the future," Diabet. Med., vol. 36, no. 8, pp. 932–941, 2019.
[11] L. H. Alotaibi et al., "Big data and diabetes: A systematic review of the literature," J. Diabetes Sci. Technol., vol. 13, no. 5, pp. 866–878, 2019.
[12] T. H. Kim et al., "Big data analysis for developing a personalized management system for diabetes mellitus," Healthcare informatics research, vol. 24, no. 2, pp. 99-107, 2018.
[13] J. C. Wylie et al., "Ethical, legal, and social considerations for incorporating machine learning into clinical decision support," Journal of Clinical and Translational Science, vol. 4, no. 5, pp. 414-421, 2020.
[14] Yang, C., Li, J., Jiang, X., & Liu, M. (2018). Privacy preserving big data analysis for personalized healthcare. IEEE Access, 6, 21862-21872.
[15] Hou, S., Wang, Y., Li, X., & Jia, Y. (2019). Privacy-preserving machine learning for diabetes prediction. Journal of medical systems, 43(10), 295.
[16] Al-Rubaiee, M., & Reddy, M. (2020). A review of privacy preserving techniques in big data analytics. IEEE Access, 8, 60111-60132.
[17] Ahmed, S., Han, J. H., & Jo, G. (2019). Security and privacy issues in big data: A survey. Journal of Network and Computer Applications, 136, 38-54.
[18] Javanmardi, M., & Mišić, J. (2020). A survey of privacy-preserving machine learning techniques. Journal of Network and Computer Applications, 155, 102602.
[19] Alotaibi, M., Alenazi, M., & Alshammari, R. (2019). Mobile health for diabetes management: Current status and future prospects. Diabetes & Metabolic Syndrome: Clinical Research & Reviews, 13(2), 1443-1450.
[20] Jin, H., & Kim, S. (2018). Wearable sensors and mobile health apps for diabetes care: A systematic review. Journal of Korean Medical Science, 33(51), e326.
[21] Wang, J., Huang, Y., & Guo, X. (2019). Telemedicine for diabetes care in China: A systematic literature review. Journal of telemedicine and telecare, 25(1), 3-13.
[22] Mridha, K., Kuri, A. C., Saha, T., Jadeja, N., Shukla, M., & Acharya, B. (2023, May). Toward Explainable Cardiovascular Disease Diagnosis: A Machine Learning Approach. In International Conference on

Data Analytics and Insights (pp. 409-419). Singapore: Springer Nature Singapore.
[23] Lu, J., Ma, S., & Sun, J. (2019). Big data analytics for diabetes management. Journal of Medical Systems, 43(6), 129.
[24] Virani, J., Daredi, N., Bhanushali, A., Shukla, M., & Shah, P. (2023, March). Mental Healthcare Analysis using Power BI & Machine Learning. In 2023 4th International Conference on Signal Processing and Communication (ICSPC) (pp. 73-76). IEEE.
[25] Zhou, X., & Chen, K. (2019). Artificial intelligence in diabetes care. Journal of Diabetes Investigation, 10(1), 24-32.
[26] Parekh, M., & Shukla, M. (2022). Survey of Streaming Clustering Algorithms in Machine Learning on Big Data Architecture. In Information and Communication Technology for Competitive Strategies (ICTCS 2021) ICT: Applications and Social Interfaces (pp. 503-514). Singapore: Springer Nature Singapore.

5

Machine Intelligence and Big Data in Diabetic Care: Laboratorian's Perspective

Arindam Ghosh[1], Aritri Bir[1], and Asitava Deb Roy[2]

[1]Department of Biochemistry, Dr. B. C. Roy Multi-Speciality Medical Research Centre, IIT Kharagpur, India
[2]Department of Pathology/Lab Medicine,
All India Institute of Medical Sciences, Deoghar, Jharkhand, India
E-mail: arindam@bcrmrc.iitkgp.ac.in; dr.aritribir@gmail.com; asitavadr@gmail.com

Abstract

The majority of our everyday activities have gone digital during the past ten years. The growing synergy of cutting-edge medical technology, innovation, and digital communication is taken into consideration by digital health. We are no longer restricted to a descriptive study of the data since machine learning allows us to detect and forecast patterns that come from inductive reasoning, which can be more valuable. The use of "what-if" models in machine learning software that explains the assumptions underlying a prediction makes it easier to determine the best course of action by understanding if and how altering particular aspects may enhance the results. Diabetes care currently faces a number of difficulties, including a shortage of diabetologists, an increase in patients, time restrictions on doctor visits, an increase in the complexity of the disease from the perspectives of clinical and patient care, difficulty meeting pertinent clinical goals, an increase in the burden of disease management on both patients and health care professionals, and health care accessibility and sustainability. Artificial intelligence and other digital technologies present a huge opportunity in this area.

According to physicists and laboratorians, artificial intelligence will make it possible to transform descriptive data into knowledge of the correlations and factors that "influence" behaviour, therefore finding the crucial elements that may enhance the predicted outcomes. Therefore, artificial intelligence has the potential to be a very useful technological instrument for helping diabetologists take full responsibility for each patient, ensuring individualised and accurate medical care. As a result, complete therapies will be able to be developed in compliance with the evidence-based standards that need to guide all therapeutic decisions.

5.1 Introduction

The prevalence of diabetes is on the rise globally, and it is projected that by 2045, approximately 700 million individuals will be affected by diabetes [1]. Diabetes mellitus, a complex and chronic metabolic disorder, affects millions of individuals worldwide. It is a chronic metabolic disorder that necessitates continuous monitoring and management, and technology has the potential to revolutionize the way the disease is managed. The management of diabetes is a multifaceted challenge, requiring precise monitoring, individualized treatment, and proactive intervention to prevent complications. In recent years, the healthcare industry has witnessed a remarkable transformation driven by advances in machine intelligence and big data analytics. These technologies have enabled healthcare providers to leverage vast amounts of data to enhance decision-making and improve patient care. This essay explores the role of laboratorians, specifically those involved in clinical diagnostics and data management, in harnessing the power of machine intelligence and big data to revolutionize diabetic care. Over the last few decades, significant advancements have been made in diabetic care, including developing new drugs, devices, and diagnostic tools. The incorporation of machine learning and big data analytics in diabetic care has transformed the way patients are managed, and their outcomes are predicted [2]. In diabetic care, laboratory medicine has also played a critical role by providing reliable and accurate test results for the diagnosis and monitoring of the disease. The aim of this chapter is to examine the potential of machine intelligence and big data analytics in diabetic care from the viewpoint of a laboratorian.

5.2 Machine Intelligence in the Field of Laboratory Science

In recent years, the integration of machine intelligence, particularly artificial intelligence (AI), has transformed laboratory operations across various scientific disciplines. Machine intelligence in the laboratory refers to the utilization

5.2 Machine Intelligence in the Field of Laboratory Science

of AI algorithms, data analytics, and automation technologies to enhance data analysis, experiment design, sample processing, and decision-making. The significant role of machine intelligence in the laboratory, focusing on its impact, applications, and future prospects is discussed below:

- Data analysis and interpretation: Machine intelligence has revolutionized data analysis in laboratory settings. Advanced AI algorithms, such as deep learning and neural networks, have been employed to process large datasets generated by high-throughput instruments and experiments. These algorithms can identify patterns, correlations, and anomalies that may be challenging for human analysts to detect. For instance, in genomics, AI-driven tools can analyze DNA sequences for variations linked to diseases, enabling faster and more accurate diagnostics [3].
- Experiment design and optimization: Machine intelligence aids in the design and optimization of experiments by considering multiple variables and constraints. Algorithms can suggest optimal conditions, reagent concentrations, and sampling intervals to maximize experimental outcomes and minimize resource consumption. This approach is particularly valuable in fields like chemistry and drug discovery, where experimentation can be costly and time-consuming [4].
- Laboratory automation: Robotics and machine learning have significantly improved laboratory automation. Automated systems, guided by AI algorithms, can perform repetitive tasks with high precision and accuracy. This reduces human error, enhances reproducibility, and accelerates data generation. In pharmaceutical research, for example, automated systems are used in high-throughput screening of drug candidates [5].
- Quality control and assurance: Machine intelligence contributes to quality control and assurance by monitoring instrument performance and data integrity. AI algorithms can identify instrument malfunctions or data inconsistencies in real time, allowing corrective actions to be taken promptly. In analytical chemistry, AI-powered systems ensure the reliability of analytical measurements [6].
- Personalized medicine: Machine intelligence plays a pivotal role in the field of personalized medicine by analyzing patient-specific data to tailor treatment plans. Laboratories use AI to interpret genetic, proteomic, and clinical data to predict disease risk, treatment response, and adverse reactions. This individualized approach improves patient outcomes and reduces adverse effects [7].

- Drug discovery and development: AI-driven drug discovery has gained prominence, with machine intelligence assisting in target identification, compound screening, and clinical trial optimization. AI algorithms analyze biological data, chemical properties, and clinical records to identify potential drug candidates, accelerating the drug development process [8].
- Biomarker discovery: Machine intelligence aids in biomarker discovery by identifying relevant molecular markers associated with diseases. Through data mining and pattern recognition, AI algorithms can pinpoint potential biomarkers, facilitating early disease detection and monitoring [9].
- Predictive analytics: Predictive analytics powered by machine intelligence is increasingly applied to laboratory management. These systems forecast equipment maintenance needs, supply chain demands, and resource allocation based on historical data and real-time inputs, ensuring efficient laboratory operations [10].
- Challenges and ethical considerations: While the benefits of machine intelligence in the laboratory are evident, it also brings challenges and ethical considerations. Data privacy, security, algorithm transparency, and bias mitigation are crucial aspects that require attention [11]. Additionally, the need for appropriate training and expertise in AI implementation within laboratory settings is essential to maximize its potential.
- Future prospects: The role of machine intelligence in the laboratory is poised to expand further. Advancements in AI, coupled with the growing availability of large datasets, will drive innovations in diagnostics, therapeutics, and scientific research [12]. Collaborative efforts among scientists, data scientists, and AI specialists will be instrumental in harnessing the full potential of machine intelligence in laboratories worldwide.

Therefore, machine intelligence is revolutionizing laboratory operations by enhancing data analysis, experiment design, automation, and decision-making processes. Its applications span various scientific domains, from genomics and drug discovery to quality control and personalized medicine. As technology continues to evolve, the laboratory's role in scientific research and healthcare will become increasingly data-driven and efficient, driven by the power of machine intelligence. In this chapter, we shall restrict ourselves to the role of machine intelligence and big data in the field of diabetic care.

5.2.1 The significance of data in diabetic care

Data lies at the heart of modern healthcare, and this is particularly true in the realm of diabetic care. Laboratorians are responsible for collecting, processing, and analyzing a myriad of data types, including blood glucose levels, HbA1c (glycated hemoglobin), lipid profiles, genetic information, and more. This comprehensive dataset provides invaluable insights into a patient's condition, enabling more precise diagnosis and the formulation of personalized treatment plans.

Laboratorians are instrumental in ensuring the accuracy and reliability of this data. They play a critical role in establishing standardized procedures for data collection and storage, ensuring the quality of laboratory tests, and maintaining the integrity of electronic health records (EHRs). As the custodians of clinical data, laboratorians are well-positioned to facilitate its efficient utilization in diabetic care.

5.2.2 The role of big data analytics in diabetic care

The integration of big data analytics in diabetic care has enabled the collection and analysis of vast amounts of patient data, including clinical, genetic, and lifestyle information. Predictive models may be created using this data to assist in identifying patients at a high risk of developing complications and allow physicians to take early action to stop or postpone the development of difficulties. For instance, big data analytics can be used to analyze the dietary and lifestyle habits of diabetic patients to identify factors that contribute to poor glycemic control [2]. Big data analytics makes it possible to create individualized treatment plans that enhance patient outcomes and save healthcare costs by identifying patient subgroups that are more likely to benefit from a certain therapy [13].

5.2.3 The role of machine intelligence in diabetic care

Machine intelligence has the potential to transform the field of diabetic care by enabling accurate and efficient diagnosis, predicting patient outcomes, and personalizing treatment plans. The use of machine learning algorithms can help identify patterns and correlations in large datasets, which can aid in disease diagnosis and risk assessment. For instance, machine learning algorithms can be used to analyze electronic health records (EHRs) of diabetic patients to identify risk factors associated with the disease and predict the likelihood of complications such as retinopathy, neuropathy, and nephropathy

[14]. Furthermore, machine intelligence can also assist in the development of personalized treatment plans based on a patient's clinical and genetic profile. For instance, machine learning algorithms can be used to predict a patient's response to a particular drug, enabling clinicians to select the most effective treatment option [15].

5.2.4 The role of laboratory medicine in diabetic care

Laboratory medicine plays a critical role in diabetic care by providing accurate and reliable test results for the diagnosis and monitoring of the disease. A variety of laboratory tests, including HbA1c, glucose, lipid profiles, and kidney function tests, are used to monitor diabetic patients and assess their response to treatment. The integration of machine intelligence and big data analytics in laboratory medicine has further enhanced the accuracy and efficiency of diagnostic testing, enabling faster and more accurate diagnosis of the disease. For instance, machine learning algorithms can be used to analyze imaging data, enabling accurate diagnosis of diabetic retinopathy, a common complication of the disease, allowing for early intervention and prevention of blindness [16].

5.2.5 Personalized diabetes management

Diabetes care has traditionally been a "one-size-fits-all" approach, with little consideration for individual differences [17]. However, with machine intelligence and big data, personalized diabetes management is becoming a reality [18]. By analyzing large amounts of data from different sources, such as electronic health records, wearables, and mobile apps, machine learning algorithms can identify patterns and correlations that are specific to each patient. This allows for tailored treatment plans that are customized to the patient's needs, preferences, and lifestyle.

One example of personalized diabetes management is the use of continuous glucose monitoring (CGM) devices. These devices measure glucose levels in real time and provide feedback to patients and healthcare providers. Machine learning algorithms can analyze the data from CGM devices and identify patterns that are specific to each patient, such as their glucose variability and insulin sensitivity [19]. This information can be used to adjust treatment plans, such as insulin dosing and timing, to optimize glycemic control [20].

5.2.6 Early detection and prevention of diabetes complications

Diabetes is a chronic disease that can lead to a range of complications, such as neuropathy, retinopathy, and nephropathy. Early detection and prevention of these complications are essential for improving patient outcomes and reducing healthcare costs [17]. Machine intelligence and big data can help in the early detection and prevention of diabetes complications by analyzing large amounts of data from different sources, such as medical imaging, laboratory tests, and electronic health records [21].

5.2.7 Predictive analytics for diabetes management

The use of statistical algorithms and machine learning methods to evaluate data and create predictions about future events is known as predictive analytics [22]. In the context of diabetes care, predictive analytics can be used to forecast a patient's risk of developing complications, such as cardiovascular disease or diabetic foot ulcers. This information can be used to guide treatment decisions and prevent or delay the onset of complications [23].

One example of using predictive analytics for diabetes management is the use of machine learning algorithms to predict hypoglycemia. Hypoglycemia, or low blood sugar, is a common complication of diabetes that can lead to seizures, coma, or even death. Machine learning algorithms can analyze data from CGM devices and other sources, such as patient behavior and medication use, to predict the likelihood of hypoglycemia. This information can be used to adjust treatment plans and prevent hypoglycemic events [24].

Predictive analytics, coupled with machine learning algorithms, has shown great potential in improving diabetes management by predicting hypoglycemia (low blood sugar) events. These predictive models help healthcare providers and patients take proactive measures to prevent severe hypoglycemic episodes [22–24]. Below, we shall discuss the use of machine learning approaches for hypoglycemia prediction and provide some insights into the results and challenges associated with these models.

1. Time series analysis: One of the primary methods for predicting hypoglycemia is through time series analysis of blood glucose data. Machine learning algorithms such as recurrent neural networks (RNNs) and long short-term memory networks (LSTMs) excel at capturing temporal dependencies in data. Time series data from continuous glucose monitors (CGMs) or self-reported data can be used to train these models.

The input features may include past blood glucose levels, insulin doses, carbohydrate intake, and physical activity. RNNs and LSTMs can learn patterns and trends in the data to predict future glucose levels, including the likelihood of hypoglycemic events.
2. Feature engineering: Feature engineering involves selecting relevant features that contribute to the prediction of hypoglycemia. These features can include meal timing, insulin type and dosage, exercise, sleep patterns, and stress levels. Machine learning models such as decision trees, random forests, and gradient boosting can be trained on these engineered features to make predictions. Feature importance analysis can help identify which factors are most influential in predicting hypoglycemia.
3. Ensemble models: Ensemble models combine multiple machine learning algorithms to improve prediction accuracy. For hypoglycemia prediction, an ensemble approach can be used to combine the strengths of different algorithms. For example, an ensemble may consist of a combination of decision trees, support vector machines, and neural networks. The final prediction is a weighted average or voting result of individual model predictions.
4. Deep learning and neural networks: Deep learning techniques, including convolutional neural networks (CNNs) and feedforward neural networks, have also been employed for hypoglycemia prediction. CNNs can be used to analyze images of food, which can help predict post-meal blood glucose levels and the risk of hypoglycemia. Feedforward neural networks can be trained on structured data like insulin dosage, carbohydrate intake, and historical blood glucose levels to make predictions.

5.2.7.1 Results and impact

Predictive analytics for hypoglycemia management has yielded promising results:

- Early warning: Machine learning models can provide early warnings of impending hypoglycemic events. Patients and healthcare providers can take preventive actions such as adjusting insulin dosages, consuming snacks, or engaging in physical activity to mitigate the risk.
- Personalized care: These models enable personalized diabetes management. Each patient's unique data profile is used to make predictions, ensuring that interventions are tailored to individual needs and patterns.

- Reduced hypoglycemia incidents: By predicting hypoglycemia in advance, patients can take proactive measures to avoid severe low blood sugar episodes, reducing the frequency of hospitalizations and emergency room visits.
- Improved quality of life: Patients experience improved quality of life as they gain more control over their condition, leading to fewer disruptions caused by hypoglycemic events.
- Resource optimization: Healthcare providers can allocate resources more efficiently. Patients at higher risk of hypoglycemia can receive more frequent monitoring and support, while those at lower risk can have fewer interventions, reducing the burden on healthcare systems.

5.2.7.2 Challenges and considerations

Despite the promising results, there are challenges associated with implementing machine learning models for hypoglycemia prediction:

- Data quality: The accuracy and reliability of predictive models heavily depend on the quality of data. Inaccurate or incomplete data can lead to unreliable predictions.
- Inter-patient variability: Diabetes is a highly heterogeneous condition. Models need to account for the variability in patient behaviors, physiology, and responses to treatment.
- Ethical considerations: The use of patient data, especially in predictive analytics, raises ethical concerns related to privacy, consent, and data security. Transparent and ethical data handling practices are essential.
- Clinical validation: Machine learning models must undergo rigorous clinical validation to ensure they are safe and effective. These models should complement clinical judgment rather than replace it.
- Continuous monitoring: Successful implementation often requires continuous monitoring through wearable devices or frequent data input from patients, which can be burdensome.
- Cost and accessibility: The adoption of machine-learning-based diabetes management may be cost-prohibitive for some individuals or healthcare systems. Ensuring accessibility is a significant challenge.

Predictive analytics powered by machine learning algorithms holds significant promise in improving hypoglycemia prediction and management in diabetes care. These models have the potential to enhance patient outcomes, reduce the risk of severe hypoglycemic events, and optimize healthcare resource allocation. However, addressing challenges related to data quality, variability among patients, ethical considerations, and clinical validation is essential for the successful integration of machine learning into diabetic care. Collaborative efforts among healthcare providers, data scientists, and patients will be crucial in harnessing the full potential of predictive analytics for diabetes management [25, 26].

5.2.8 Personalized treatment plans

Traditionally, diabetes care has followed a standardized or "one-size-fits-all" model, where patients with diabetes are provided with a uniform set of guidelines, medications, and recommendations for managing their condition. However, this approach overlooks the significant individual differences among patients with diabetes, including factors such as age, genetics, lifestyle, and the specific type and stage of diabetes. Here is an analysis of why this approach is inadequate and how a personalized approach can improve diabetes care:

- Variability in diabetes types: Diabetes is not a single condition but a group of diseases with distinct causes and characteristics, including type-1 diabetes, type-2 diabetes, gestational diabetes, and others. Each type requires different management strategies.
- Genetic variations: People have unique genetic backgrounds that influence how they respond to treatments and medications. For example, some individuals may have genetic predispositions that make them more resistant to certain medications or more prone to side effects.
- Lifestyle and behavior: Lifestyle factors such as diet, physical activity, stress levels, and sleep patterns play a critical role in diabetes management. These factors vary widely among individuals and should be considered in treatment plans.
- Comorbidities: Many individuals with diabetes also have other health conditions, such as hypertension, cardiovascular disease, or kidney disease. These comorbidities require tailored treatment approaches.
- Response to medications: People can react differently to the same medication. Factors like metabolism, kidney function, and interactions with other drugs need to be taken into account.
- Blood glucose patterns: Continuous glucose monitoring and analysis of blood glucose patterns can provide valuable insights into an individual's unique response to foods, medications, and other factors.
- Psychosocial factors: Emotional well-being, mental health, and social support can impact a person's ability to manage their diabetes effectively. These factors should be addressed individually.

5.2 Machine Intelligence in the Field of Laboratory Science

To transition from the traditional "one-size-fits-all" approach to a personalized diabetes care model, the following innovations and techniques can be implemented:

- Genetic profiling: Genetic information is increasingly being integrated into the management of diabetes. Laboratorians can assist in genetic profiling to identify specific gene variants associated with diabetes susceptibility, drug metabolism, and response to treatment. This information can guide clinicians in selecting the most appropriate therapies and predicting a patient's likelihood of developing complications.
- Continuous glucose monitoring: Continuous glucose monitoring (CGM) is another example of personalization in diabetic care. CGM devices provide real-time data on blood glucose levels, offering a more comprehensive view of glycemic control compared to intermittent fingerstick measurements. Laboratorians play a role in ensuring the accuracy and reliability of CGM systems and interpreting the data they generate.
- Medication optimization: Medication management in diabetes is a complex task. Different individuals may respond differently to the same medication, and factors like metabolism, kidney function, and potential drug interactions must be considered. Laboratorians can collaborate with pharmacists and clinicians to optimize medication regimens, taking into account individual patient data.
- Artificial intelligence and machine learning: Use AI algorithms to analyze patient data and provide personalized recommendations for insulin dosages, diet, and lifestyle modifications.
- Telehealth and remote monitoring: Enable patients to access healthcare professionals remotely for regular check-ins, reducing the need for in-person visits and improving access to care.
- Patient education: Provide patients with personalized educational materials and support that address their unique needs and challenges.
- Shared decision-making: Involve patients in the decision-making process, considering their preferences and goals when creating treatment plans.
- Behavioral support: Offer resources and counseling to help patients make sustainable lifestyle changes.
- Interdisciplinary care teams: Collaborate with a multidisciplinary team of healthcare professionals, including dietitians, psychologists, and pharmacists, to provide holistic care.

118 *Machine Intelligence and Big Data in Diabetic Care*

By embracing these innovations and techniques, healthcare providers can shift from a one-size-fits-all model to a personalized approach, ultimately improving diabetes management and outcomes for individuals with diabetes. The workflow is schematically represented in Figure 5.1.

Personalization is a central theme in contemporary healthcare, and it is particularly relevant in diabetic care. The heterogeneity of diabetes requires treatment strategies that are adapted to the unique needs and characteristics of each patient. Laboratorians play a key role in this process by providing the data and insights needed to tailor treatment plans.

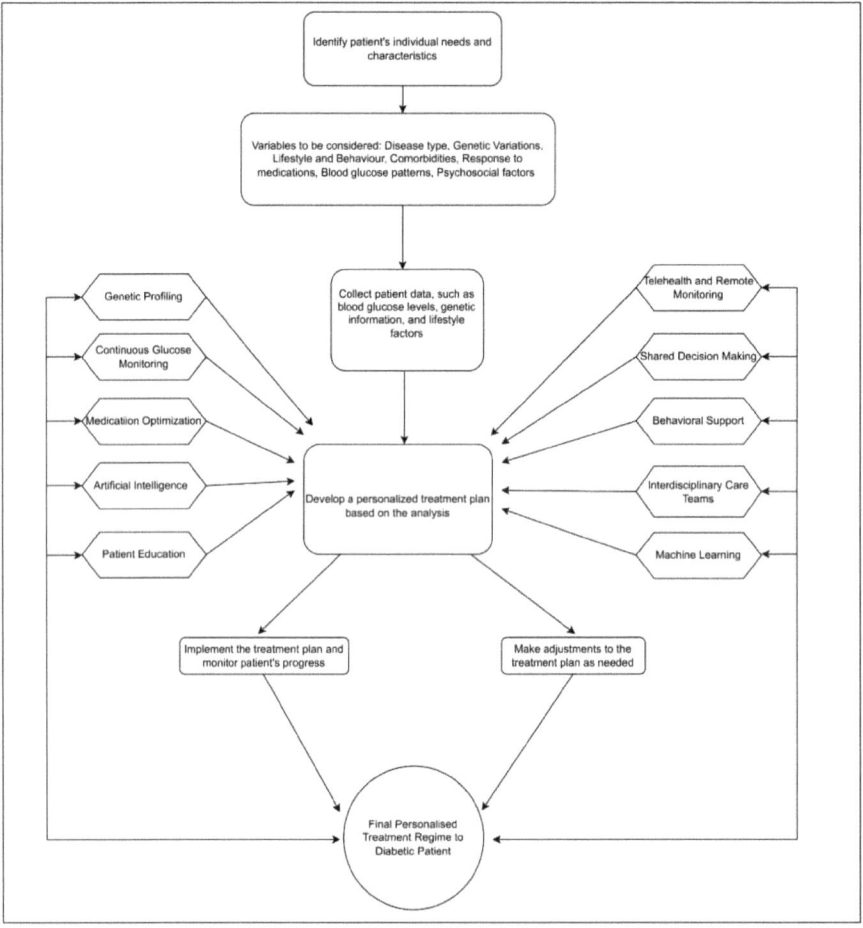

Figure 5.1 The workflow of constructing personalized diabetes treatment regime.

5.2.9 Remote monitoring and telemedicine

The advent of telemedicine and remote monitoring technologies has been accelerated by the COVID-19 pandemic. Laboratorians are well-placed to facilitate remote monitoring of diabetic patients. They ensure the seamless transmission of data between patients and healthcare providers, allowing for real-time tracking of vital parameters such as blood glucose levels, medication adherence, and lifestyle factors.

Remote monitoring enhances patient engagement and enables timely adjustments to treatment plans. Patients can receive personalized guidance and interventions, reducing the need for frequent in-person visits, particularly important for those living in remote or underserved areas. The role of laboratorians in ensuring the reliability of remote monitoring systems cannot be overstated, as data accuracy is paramount in remote care.

5.2.10 Challenges and opportunities

While machine intelligence and big data analytics present huge opportunities in diabetic care, some challenges must be addressed.

1. Manpower: One of the main challenges is the shortage of diabetologists, which can lead to delays in diagnosis and treatment. The use of machine learning algorithms and big data analytics can help overcome this challenge by enabling early identification of the disease and personalized treatment plans that reduce the burden on healthcare professionals.
2. Disease complexity: Another challenge is the increasing complexity of the disease from the perspectives of clinical and patient care, which makes it difficult to meet pertinent clinical goals. Machine intelligence and big data analytics can help overcome this challenge by providing clinicians with real-time data on patient health, enabling them to make evidence-based decisions that improve patient outcomes.
3. Data privacy and security: Protecting patient privacy is of paramount importance. Laboratorians must implement robust data security measures to safeguard sensitive health information. This includes encryption, access controls, and compliance with regulatory frameworks such as the Health Insurance Portability and Accountability Act (HIPAA) in the United States.
4. Informed consent: Collecting and using patient data for research and analysis requires informed consent. Laboratorians must ensure that patients are fully informed about the purposes and potential risks of data

usage, and that they provide their consent voluntarily. This is particularly relevant when utilizing data for AI and ML model training.
5. Data bias and fairness: Machine learning algorithms can inherit biases present in the data they are trained on. Laboratorians should be vigilant in identifying and mitigating biases to ensure that AI systems provide fair and equitable care to all patient populations.
6. Algorithm transparency and explainability: The "black box" nature of some AI algorithms can be a challenge in healthcare. Laboratorians should work to develop transparent and explainable AI models, enabling clinicians and patients to understand the rationale behind recommendations and decisions.
7. Data quality: Ensuring the quality of data is a perpetual challenge. Laboratorians must implement data validation and quality assurance processes to detect and rectify errors, inconsistencies, and missing data that can adversely affect the accuracy of AI algorithms.

5.3 Conclusion

The integration of machine intelligence and big data analytics in diabetic care has the potential to transform the field of diabetic care by enabling accurate and efficient diagnosis, predicting patient outcomes, and personalizing treatment plans. However, the adoption of these technologies requires overcoming several challenges, such as data quality, privacy concerns, and regulatory issues. Collaboration between healthcare providers, researchers, and technology companies is essential for realizing the full potential of machine intelligence and big data in diabetes care. Laboratory medicine plays a critical role in diabetic care by providing accurate and reliable test results for the diagnosis and monitoring of the disease. Combining machine intelligence, big data analytics, and laboratory medicine can improve patient outcomes, lower healthcare costs, and improve the overall quality of care for diabetic patients [27–30].

References

[1] Saeedi P, Petersohn I, Salpea P, et al. Global and regional diabetes prevalence estimates for 2019 and projections for 2030 and 2045: Results from the International Diabetes Federation Diabetes Atlas, 9th edition. *Diabetes Res Clin Pract.* 2019;157:107843. doi:10.1016/j.diabres.2019.107843

[2] Kharroubi SA, Herman WH. Big Data and Machine Learning in Diabetes Research: Recent Advances and Future Directions. Diabetes Care. 2019; 42(8):1436-1444.

[3] Hannigan, G. D., et al. (2018). Diagnostic Potential of the Gut Microbiome and Metabolome in Patients with Early-Stage Pancreatic Cancer. Cancer Research, 78(16), 3871-3881.

[4] Schmidt, T., et al. (2019). Realizing the Potential of High-Throughput Experimentation in Materials Discovery. MRS Bulletin, 44(7), 509-514.

[5] Simm, J., et al. (2019). Data-Driven Hypothesis Weighting Increases Prediction Accuracy in Experimental Biophysical Screening. Biotechnology Journal, 14(1), 1800080.

[6] Cristobal, G., et al. (2019). Artificial Intelligence for Analytical Chemistry—A Review. Analytica Chimica Acta, 1099, 1-17.

[7] Ehteshami Bejnordi, B., et al. (2020). Diagnostic Assessment of Deep Learning Algorithms for Detection of Lymph Node Metastases in Women with Breast Cancer. JAMA, 318(22), 2199-2210.

[8] Lima, A. N., et al. (2020). Machine Learning in Drug Discovery and Development. Computational and Structural Biotechnology Journal, 18, 241-246.

[9] Rodriguez-Esteban, R. (2021). Machine Learning Applications in Drug Development. Computational and Structural Biotechnology Journal, 19, 654-662.

[10] Lowe, H. J., et al. (2017). The Challenges of Implementing Machine Learning Algorithms in Clinical Medicine. Journal of the American Medical Informatics Association, 25(6), 631-635.

[11] Lorenzi, N. M., et al. (2020). Ethical Considerations and the Approval of Machine Learning-Based Medical Devices. Science Translational Medicine, 12(538), eaaz2564.

[12] Angermueller, C., et al. (2016). Deep Learning for Computational Biology. Molecular Systems Biology, 12(7), 878.

[13] Raghupathi W, Raghupathi V. Big data analytics in healthcare: promise and potential. Health Inf Sci Syst. 2014 Feb 7;2:3. doi: 10.1186/2047-2501-2-3. PMID: 25825667; PMCID: PMC4341817.

[14] Luo G, Stone BL, Johnson MD, et al. Identifying Risk Factors for Diabetic Complications Using Electronic Health Records. Journal of Medical Systems. 2015;39(7):75.

[15] Zou Q, Qu K, Luo Y, Yin D, Ju Y, Tang H. Predicting Diabetes Mellitus With Machine Learning Techniques. Front Genet. 2018 Nov

6;9:515. doi: 10.3389/fgene.2018.00515. PMID: 30459809; PMCID: PMC6232260.

[16] Ting DSW, Cheung CY, Lim G, et al. Development and Validation of a Deep Learning System for Diabetic Retinopathy and Related Eye Diseases Using Retinal Images From Multiethnic Populations With Diabetes. JAMA. 2017;318(22):2211-2223.

[17] American Diabetes Association. (2021). Standards of medical care in diabetes—2021 abridged for primary care providers. Clinical Diabetes, 39(1), 14-26. doi: 10.2337/cd20-as01

[18] Basu, A., Dube, S., Veettil, S., & Slama, M. (2020). Use of machine learning in diabetes management. Current Diabetes Reports, 20(11), 1-9. doi: 10.1007/s11892-020-01418-y

[19] Emam, K., Jonker, E., & Moreau, K. (2019). Machine learning for diabetes decision support: A systematic review. Journal of Diabetes Science and Technology, 13(4), 754-766. doi: 10.1177/1932296819851934

[20] Mazzeo, J., Hripcsak, G., & Bian, J. (2020). Machine learning for personalized diabetes management: A review. Journal of Diabetes Science and Technology, 14(1), 98-106. doi: 10.1177/1932296819875542

[21] Zhu, H., Wang, X., & Tao, D. (2018). Big data analytics in healthcare. In Healthcare Data Analytics (pp. 3-23). Springer, Cham. doi: 10.1007/978-3-319-65753-9_1

[22] Ali, I., Naeem, M. A., Siddiquei, M. M., & Khan, S. A. (2021). Predictive analytics in healthcare: A review. International Journal of Advanced Computer Science and Applications, 12(2), 186-195. doi: 10.14569/ijacsa.2021.0120219

[23] Chen, I. Y., & Asch, S. M. (2017). Machine learning and prediction in medicine—beyond the peak of inflated expectations. New England Journal of Medicine, 376(26), 2507-2509. doi: 10.1056/nejmp1702071

[24] Ludvigsson, J., Lebre, M. A., & Gudbjörnsdottir, S. (2020). Risk prediction of severe hypoglycemia in people with type 1 diabetes using machine learning and data from the HARPdoc study. Diabetes Technology & Therapeutics, 22(12), 917-925. doi: 10.1089/dia.2020.0142

[25] Mohapatra, Chinmayee, et al. "Usage of Big Data prediction techniques for predictive analysis in HIV/AIDS." Big Data Analytics in HIV/AIDS Research. IGI Global, 2018. 54-80.

[26] Mishra, Sushruta, Hrudaya Kumar Tripathy, and Biswa Acharya. "A precise analysis of deep learning for medical image processing." Bio-inspired neurocomputing (2021): 25-41

[27] Mohanty, Cheena, et al. "Using Deep Learning Architectures for Detection and Classification of Diabetic Retinopathy." Sensors 23.12 (2023): 5726.
[28] R. Janapati, U. Desai, S. A. Kulkarni, and S. Tayal, Human-Machine Interface Technology Advancements and Applications. CRC Press, 2023.
[29] Ankit Vijayvargiya, Bharat Singh, Rajesh Kumar, Usha Desai, Jude Hemanth, "Hybrid Deep Learning Approaches for sEMG Signal-Based Lower Limb Activity Recognition", Mathematical Problems in Engineering, vol. 2022, Article ID 3321810, 12 pages, 2022.
[30] U. Desai, C. G. Nayak and G. Seshikala, "An efficient technique for automated diagnosis of cardiac rhythms using electrocardiogram," 2016 IEEE International Conference on Recent Trends in Electronics, Information & Communication Technology (RTEICT), Bangalore, India, 2016, pp. 5–8.

6

EfficientNetB3-DTL: Classification of Diabetic Retinopathy Images using Modified EfficientNetB3 with Deep Transfer Learning

Ch.Rajendra Prasad[1], Sreedhar Kollem[1], Srinivas Samala[1], B. Srinivas[2], Ravichander Janapati[1], and Srikanth Yalabaka[1]

[1]Department of ECE, SR University, India
[2]Department of Computer Science & Engineering (Networks), KITS, India
E-mail: chrprasad20@gmail.com; ksreedhar829@gmail.com; srinu486@gmail.com; bs.csn@kitsw.ac.in; chander3818@gmail.com; srikanthyelabaka7131@gmail.com

Abstract

Diabetes causes diabetic retinopathy (DR), a medical disease. How long a person has had diabetes is a major factor in whether or not they develop retinopathy. Due to damage of the retinal blood vessels, there may be no first symptoms or just a small visual impairment. Eventually, it may result in blindness. Identifying the first clinical indicators of DR is crucial for intervening and successfully treating the disorder. To prevent irreversible vision loss, frequent eye exams are required to lead the individual to a doctor for a full ocular examination and treatment as soon as feasible. Yet, because of limited resources, screening is not possible. Thus, developing technologies, such as deep learning, for the automated identification and categorization of DR are alternate screening approaches, thereby reducing the cost of the system. In recent years, people have been working on methods based on deep learning to identify and interpret DR. This chapter presents a classification of diabetic retinopathy images using modified EfficientNetB3 with deep transfer learning (DTL). The proposed EfficientNetB3-DTL model employs a dataset from Kaggle. The data is pre-processed and augmented before applying it to the transfer learning model. The pre-trained EfficientNetB3 model is employed as a DTL model by modifying the layers in the basic architecture. For training 80% and for testing 20% of data is used. The performance proposed DTL model is evaluated in terms of accuracy, sensitivity, and specificity with Adam optimizer. The results show that the proposed DTL model achieves better accuracy as compared to conventional models.

Keywords: Deep transfer learning, data processing, diabetic retinopathy, EfficientNetB3, Adam optimizer, EfficientNetB3-DTL.

6.1 Introduction

To improve healthcare outcomes, it is essential to detect diseases at an early stage. Diabetes, a metabolic disorder that results in increased glucose levels in the bloodstream due to inadequate insulin production, is a widespread condition that affects approximately 425 million adults globally. Diabetes can have severe consequences, including damage to the heart, kidneys, nerves, and eyes. Diabetic retinopathy (DR) is a complication of diabetes that can result in visual impairment if left untreated. It is characterized by the dilation and leakage of retinal blood vessels. Approximately 2.6% of all occurrences of blindness worldwide can be attributed to it. Patients with long-standing diabetes are more likely to develop DR. Regular retinal screenings are essential for diabetic patients because of the importance of detecting DR early and treating it to prevent blindness [1]. The detection of DR, which manifests as a variety of lesions including soft and hard exudates (EX), microaneurysms (MA), and hemorrhages (HM), is done using retinal images [2].

The initial sign of DR is the presence of a microaneurysm (MA), which appears as a red dot on the retina and has a diameter of less than 125 m and sharp borders because of compromised vessel walls. Larger, asymmetrical patches on the retina (greater than 125 m in size) are hemorrhages (HM). Flame HM has more surface, while blot HM is more in-depth. Plasma leakage causes the retina to develop bright yellow patches known as hard exudates. They can be seen in the outer layers of the retina and are distinguished by their sharp borders. Cotton-like white patches, or soft exudates, appear on the retina when nerve fibers swell. They often come in a circular or oval form. Red lesions are those caused by MA and HM, while bright lesions are those caused by soft and hard exudates (EX). DR is classified into five categories based on the severity of the lesions; they are absence of DR, mild DR, moderate DR, severe DR, and proliferative DR.

Automatic DR detection approaches save both money and time compared to human diagnosis. Compared to automated procedures, manual diagnosis is more laborious and prone to error. To identify and classify DR, our research focuses on state-of-the-art AI-driven algorithms that make use of deep learning [28–30]. Adam, SGDM, and RMSProp are used to evaluate DTL model correctness, sensitivity, and specificity.

6.2 Related Work

The authors of the paper [3] introduced a deep learning framework that makes use of a convolutional neural network (CNN) with three separate layers to recognize different retinal layers and hence aid in the prediction of disorders such as DR, drusen, choroidal neovascularization, and DME. Optical coherence tomography (OCT) pictures were optimized by removing background noise using a number of pre-processing methods to increase image quality. The end outcome was supposed to be something accurate and trustworthy. Using deep transfer learning (DTL) strategy with Inception-v3 network, Li, Feng, et al. created an autonomous diabetic retinopathy detection system that works with color fundus images of the retina. According to the results of his research [4], the accuracy of their classification approach was 93.49%, with a sensitivity of 96.93% and a specificity of 93.45%.

The ultrawide-field fundus pictures presented in [5] provided high-resolution imaging of the retina's surface. The research centered on the use of deep learning algorithms, namely the 34-layer architecture ResNet-34, for the autonomous detection of diabetic retinopathy. To further improve outcomes, the ETDRS-7SF (early treatment of DR study-7 standard field) image segmentation approach was added and the results validated with a number of different metrics. A CNN method for identifying retinopathy was created by Samanta, Abhishek, et al., and tested on moderate-sized datasets. They used the DENSENET121 architecture [6], which had already been trained on the ImageNet dataset. According to their findings, this model can distinguish between healthy and unhealthy eyes with an F1 score of 0.97.

CNN models such as AlexNet, VGG-16, and SqueezeNet were introduced in [7] as part of a pre-trained image classification technique for diabetic retinopathy. To streamline the learning curve for the image datasets, they developed a fully connected neural network classifier based on a specific five-layer CNN model. They used ReLU activation function to boost performance. Classifying images of diabetic retinopathy was the focus of the research presented in [8], which made use of the Messidor dataset, which included

1200 color fundus images. The study used pre-trained CNN networks to attain accuracies of 92.38% for the AlexNet, 93.7% for the VGG-16, and 94.4% for the SqueezeNet.

Microaneurysm analysis was studied in order to detect DR and DME in their earliest stages, as detailed in the cited study [9]. An advanced model of deep convolutional neural networks (DCNNs) was used to achieve this and a lesion identification technique using fundus pictures. Ophthalmologists were able to distinguish between proliferative and non-proliferative forms of diabetic retinopathy by using a semantic segmentation approach to classify the images. Researchers were able to improve the accuracy and efficiency with which they diagnosed various illnesses because of this strategy. The authors presented a methodology in [10] that utilized pre-trained CNN models, namely AlexNet, VGG16, and Inception Net V3, for the purpose of classifying a dataset consisting of 166 images sourced from Kaggle. The method that was put forward attained accuracies of 37.43%, 50.03%, and 63.23%, correspondingly.

The authors of [11] suggested an entropy-based enhancement strategy to enhance the properties of images in a diabetic retinopathy (DR) dataset with uneven distribution. The increased classification performance was achieved by the use of a hybrid neural network and a low computing cost methodology. In [12], Gargeya Rishab and Theodore Leng presented a deep neural network (DNN)-based data-driven solution for the automatic detection of DR. When evaluated on a local dataset, their proposed model had a 94% sensitivity and a 98% specificity for AUC 0.97. In addition, on the MESSIDOR 2 and E-Ophtha datasets, the suggested model attained an AUC of 0.94 and 0.95, respectively. The goal of this research [13] was to develop a model of adaptive histogram equalization with contrast constraints for diabetic retinopathy (DR) image segmentation with the aim of improving image quality. The image was augmented by employing a Bayesian optimization method, and the evaluation findings were enhanced through adjustments to the inception-v4 model's hyperparameters.

A CNN-based method for diabetic retinopathy was proposed by Harry Pratt [14]. Over 70% accuracy was achieved across the entire dataset after training the model with SGD at a learning rate of 0.0001–0.0003. Accuracy was 30%, sensitivity was 95%, and specificity was 75% for the trained network. Binocular networks using convolutional neural networks (CNN) to record fundus images from both eyes were proposed in a Siamese-like topology [15]. The network's characteristics were modified such that transfer learning could be used for categorization and forecasting. The quadratic kappa score was calculated as one of the metrics used to assess the model's efficacy in this study. DCNN are presented in [16] for the identification of diabetic retinopathy (DR). In this work, DR classification is accomplished using CNN in conjunction with a gradient boosting tree (GBT). The network is trained to look for and use features such as microaneurysms, blood vessel recognition, and the presence of hard exudates and red lesions. For example, the accuracy for hard exudates + GBT was 89.4%, for red lesions + GBT it was 88.7%, for microaneurysms + GBT it was 86.2%, for blood vessels + GBT it was 79.1%, and for CNN it was 91.5 % without data pre-processing and 94.5% with data pre-processing.

Automatic methods for determining the severity of retinal disorders from photographs of the retina were proposed in a study [17]. The suggested method used the EYEPACS dataset for training and testing a machine learning algorithm based on the inception network architecture for classifying fundus images. The accuracy of the data achieved by utilizing these methods was higher than that of more conventional procedures. According to reference [18], Valentina Bellemo conducted a study on the screening of diabetic retinopathy in Africa. The study employed two CNN architectures, namely VGGNet and ResNet, to classify 4504 retina photographs sourced from a publicly accessible dataset. The proposed methodology exhibited a substantial area under the curve (AUC) concerning referral DR, achieving a sensitivity of 92.25% and a specificity of 89.04%. Furthermore, the technique that was suggested exhibited a sensitivity of 99.42% for diabetic retinopathy, which poses a risk to vision and a sensitivity of 97.19% for diabetic macular edema.

The following are the main contributions of the chapter:

- Pre-process the data by random brightness, contrast and scale down the images for the EffientNetB3.
- Propose a multi-class DR classification through DTL, i.e., EffientNetB3-Deep Transfer Learning (EfficientNetB3-DTL). Produce the final prediction probabilities, by determining the suitable number of layers experimentally in EffientNetB3-DTL.
- Test the proposed EfficientNetB3-DTL model using diabetic retinopathy (DR) dataset to evaluate its performance.

6.3 Proposed Model

In this section, we introduce the EfficientNetB3-DTL model for image classification of diabetic retinopathy. The suggested architecture of the

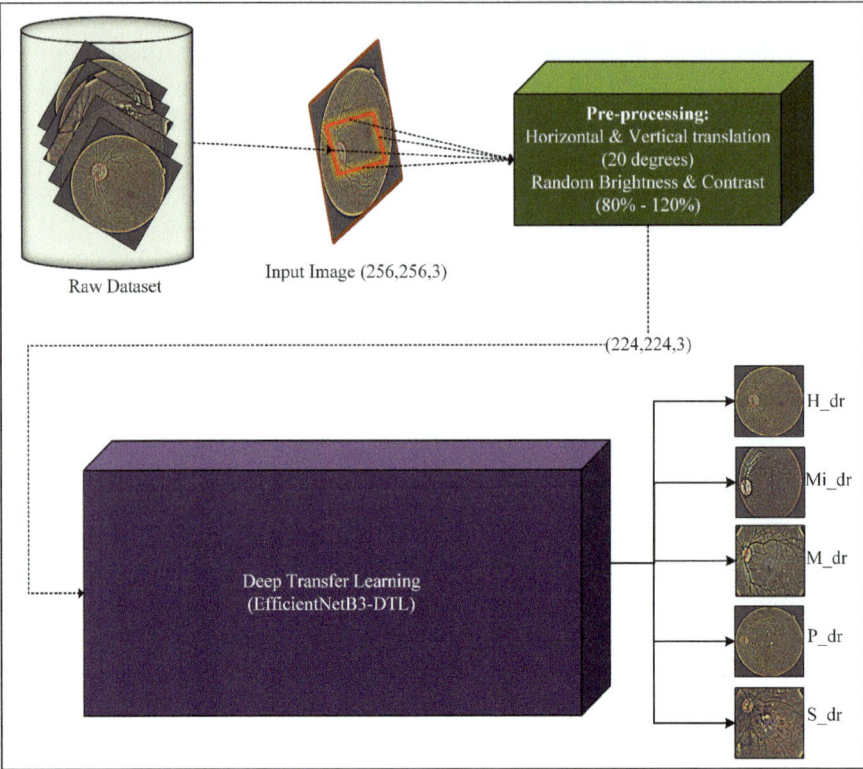

Figure 6.1 The architecture of the proposed DR classification through DTL.

EfficientNetB3-DTL model is depicted in Figure 6.1. The suggested approach involves applying EfficientNetB3-DTL on pre-processed input data. For the purposes of feature extraction and classification, DR pictures are processed with EffientNetB3.

6.3.1 Dataset

For experimentation of the proposed system, the diabetic retinopathy (DR) dataset is collected from the Kaggle database. This DR dataset consists of five classes, namely healthy DR (H_dr), mild DR (Mi_dr), moderate DR (M_dr), proliferative DR (P_dr), and severe DR (S_dr). The sample images in each class of the DR dataset are shown in Figure 6.2. The images of each class are having the same size of $256 \times 256 \times 3$. The number of images in each class of DR dataset is illustrated in Figure 6.3.

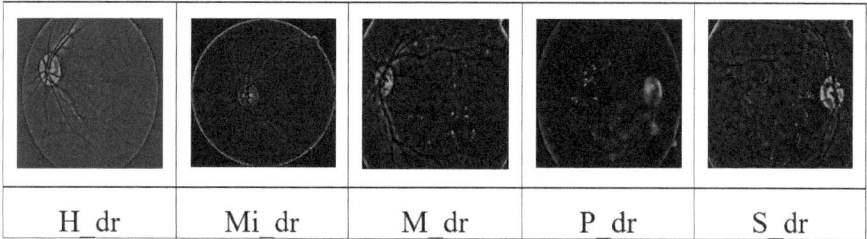

Figure 6.2 Sample images from the dataset.

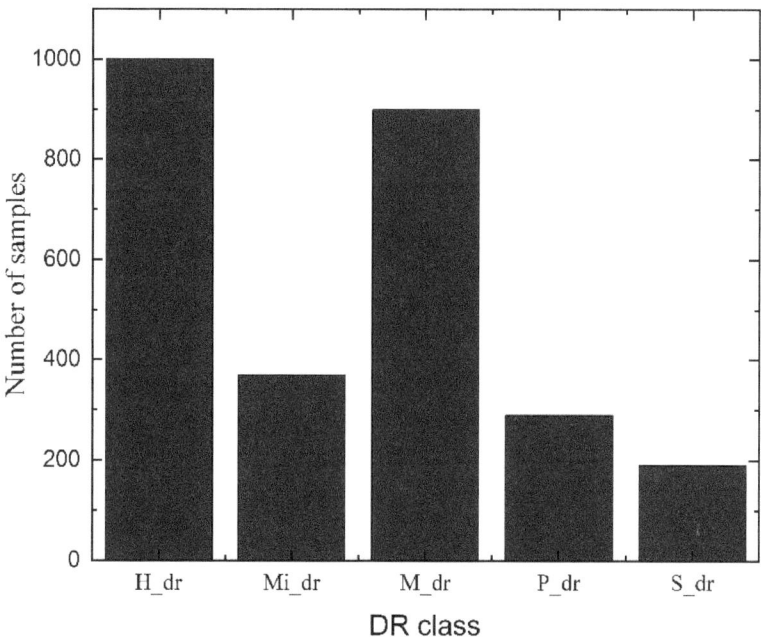

Figure 6.3 DR dataset samples in each class.

6.3.2 Data pre-processing

The DR dataset has an unequal number of samples in each class. The unequal sample size may affect the performance of the model; hence in the proposed model, 150 images in each class are considered. Before applying to the transfer learning model, the data need to be pre-processed according to the requirement of the transfer learning model. In the proposed model, EffientNetB3 is employed as a transfer learning model. An input image size of $224 \times 224 \times 3$ is needed by the EffientNetB3 model. However, the DR dataset only contains $256 \times 256 \times 3$ pixel images. In the data pre-processing

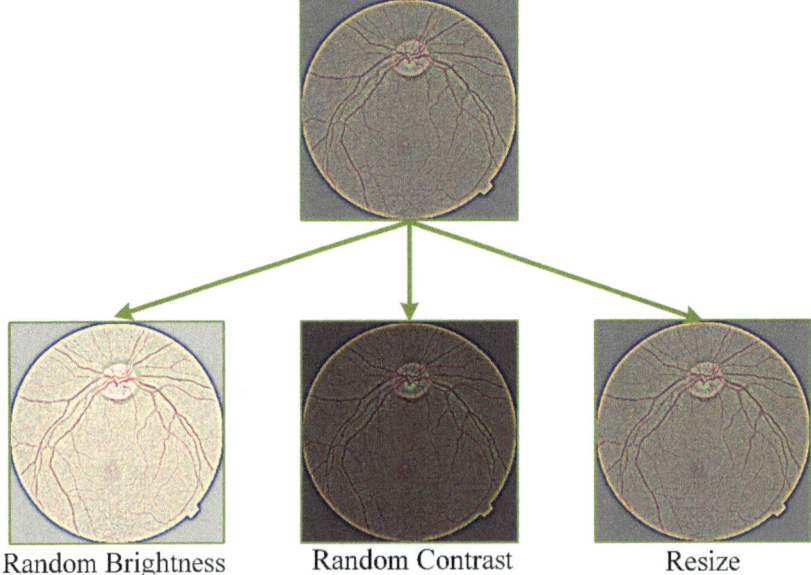

Figure 6.4 Pre-processing of H_dr image.

images are resized and, in addition, random brightness and contrast are varied by ±20% from the original image H_dr is depicted in Figure 6.4.

6.3.3 Deep transfer learning

Learning a new activity more efficiently by drawing on prior experience with a similar task is an example of transfer learning. Whenever a deep learning network's target dataset does not have enough labeled data, it uses a technique called transfer learning. The time and energy required to develop a system that uses deep learning techniques and then train it on a massive dataset is substantial. Rapid prototyping, efficient use of resources, and top-notch performance are all aided by transfer learning. In recent times, for image processing and classification issues in computers, the pre-trained foundational models, like AlexNet, ResNet50, VGG 16, VGG 19, MobileNet, EfficientNet, and InceptionV3, are most commonly employed [20]. In [21], the authors showed seven pre-trained models created with the EfficientNet framework and experimented with various combinations of depth and width for their CNN models, namely EfficientNetB0 to EfficientNetB7. These pre-trained transfer learning models provide better classification accuracy with fewer parameters. In addition, when the ImageNet dataset [22] was applied to these EfficientNet pre-trained models, in comparison to any earlier models, they excel both in precision and in the amount of parameters required for modeling.

EfficientNetB3: Models based on the efficient tradeoff between computing needs and accuracy can be found in EfficientNet. In this work, EfficientNet-B3 [23] transfer learning model is employed and illustrated in Figure 6.5. The first convolutional layer's neurons focus on a localized region of the image to extract relevant information. This space will be equal in size to the filters employed, assuming a 3×3 array. This is the neuron's receptive field. The second CNN layer convolved each neuron with the same 3×3 area as the one before it, but the field of reception is far larger than the source image. Each neuron in the network has a progressively bigger receptive field the further we travel into the network.

In the EfficientNet transfer learning model, the MBConv layer is fundamentally constructed using mobile inverted bottleneck convolutions. This MBConv is borrowed from the MobileNet transfer learning models [24]. The MBConv uses depthwise separable convolutions, which are made up of two convolution layers: one depthwise and one pointwise, stacked on top of one another. Then, two further concepts – inverted residual connections or blocks and linear bottlenecks are taken from a second upgraded version of MobileNet (MobileNet-V2).

The first concept in MobileNetV2 is the inverted residual blocks (IRB) as shown in Figure 6.6. The utilization of skip connections amidst layers that possess a substantial quantity of channels (64) is present in the initial residual blocks created from ResNets [25] in Figure 6.6 (a). The residual block reduces the total number of channels to 16, resulting in a decrease in the required parameters for the subsequent layer's 3×3 convolutions. In the IRB seen in Figure 6.6 (b), the size of the connected channels is inverted, resulting in skip connections between the smaller layer with fewer channels. Therefore, we refer to them as IRBs. The IRBs employ depthwise separable convolutions; hence, it has fewer parameters than the original ResNet residual block, even if the layer's internal channel count rises to 64.

The layer Conv (1,16) highlighted in bold in Figure 6.6 (b) features a linear activation function, evidencing MobileNetV2's second key idea, linear bottlenecks. In this layer at these locations of the network channels are being reduced. Therefore, this layer is classified as a bottleneck layer. For IRBs, commonly used in the ReLU activation function due to its inability to handle negative numbers, CNN designs are unsuitable. However, the layer with fewer channels is outperformed by the linear activation function.

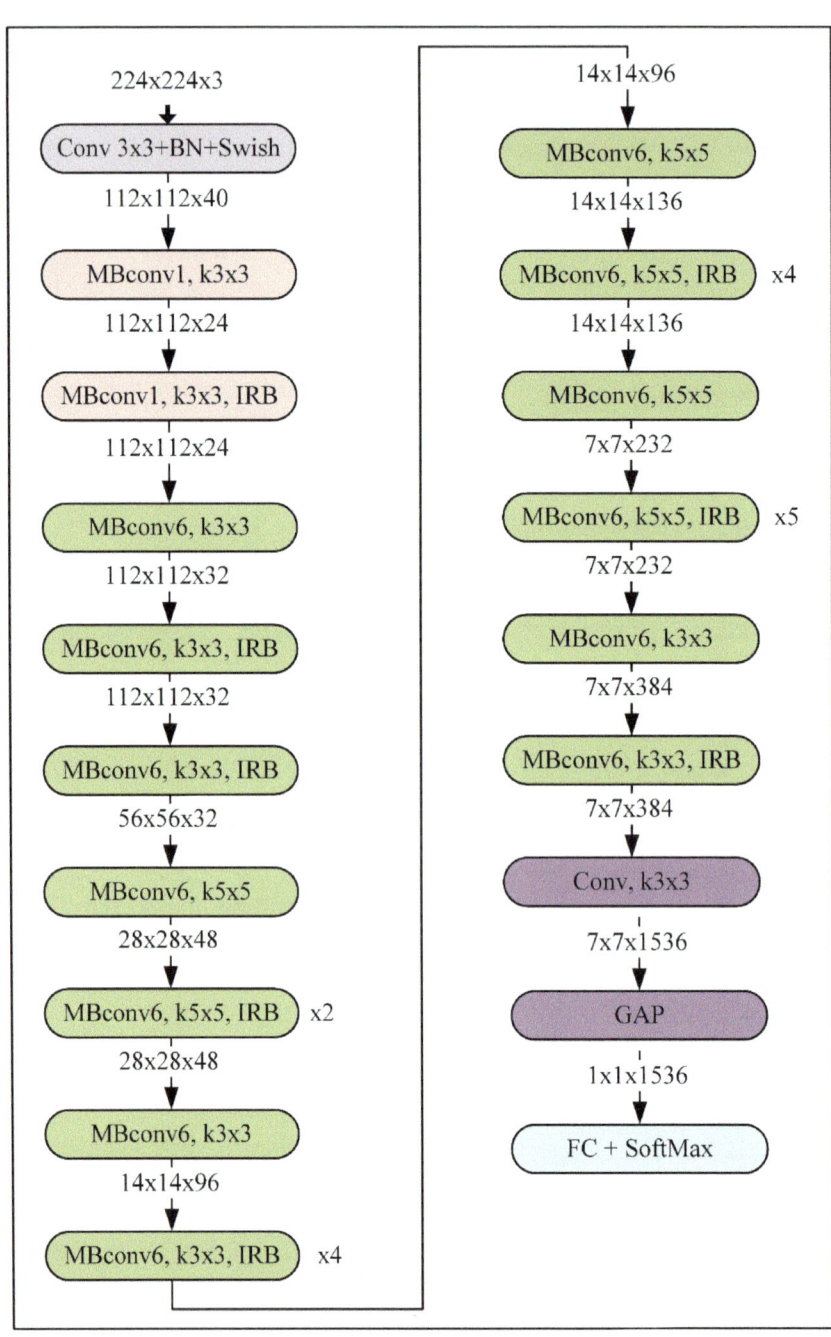

Figure 6.5 The proposed EfficientnetB3 DTL structure and its associated parameters.

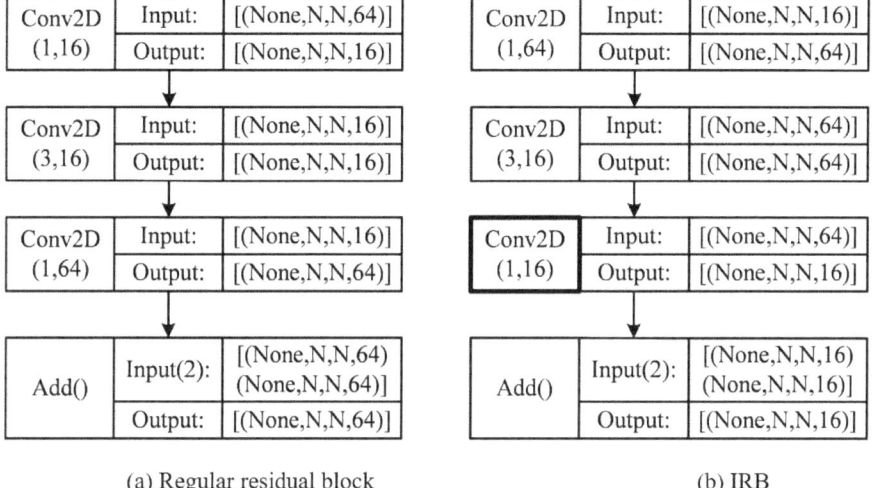

(a) Regular residual block   (b) IRB

Figure 6.6 Example of IRB.

Moreover, an alternative activation function to ReLU is proposed, a new activation function called swish is employed in this network. In contrast to the ReLU and LeakyReLU functions, the swish activation function provides a smoother activation, as illustrated in Figure 6.7. This streamlined design has similar performance benefits. Swish can be defined formally as

$$f_{\text{swish}}(x) = 2x\sigma(\beta x) = \begin{cases} \beta = 0 & \text{for} \quad f_{\text{swish}}(x) = x \\ \beta \to \infty & \text{for} \quad f_{\text{swish}}(x) = 2\max(0, x) \end{cases}, \quad (6.1)$$

where $\beta \geq 0$ is a training-dependent parameter that can be learned by the CNN model. If $\beta = 0$, then f_{Swish} turn into the linear activation function and as β tends ∞, then f_{Swish} is similar to the ReLU function, but with a smoother curve, as shown in Figure 6.7.

Finally, the features retrieved throughout the CNN model are pooled together to form a single vector. The pooling process in the proposed model uses global average pooling (GAP). In lieu of incorporating fully connected layers onto the feature maps, the global average pooling (GAP) technique computes the mean of each feature map. The resulting vector is subsequently inputted into the SoftMax layer, as depicted in Figure 6.8.

Global average pooling outperforms fully connected layers by promoting strong feature map-category correspondences. Since it has no tunable parameters, global average pooling helps prevent overfitting, which is a key

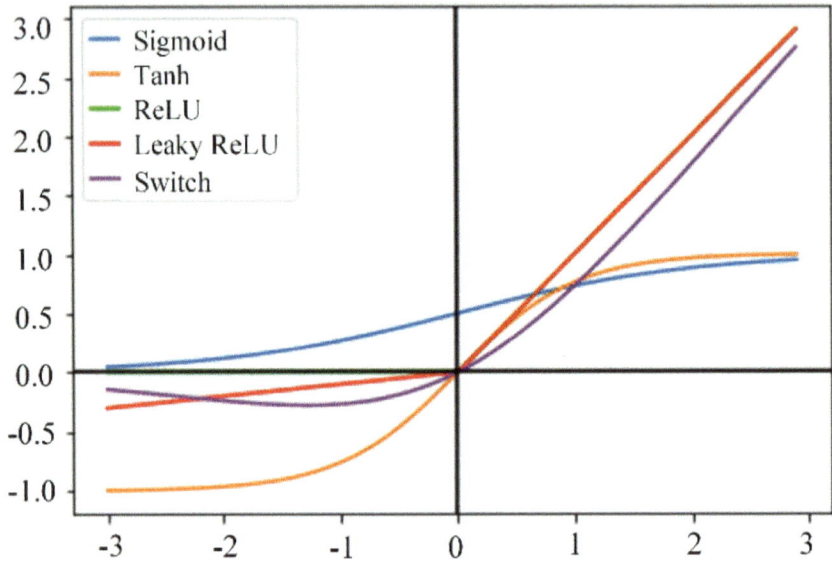

Figure 6.7 Comparison of swish activation function with other activation functions [26].

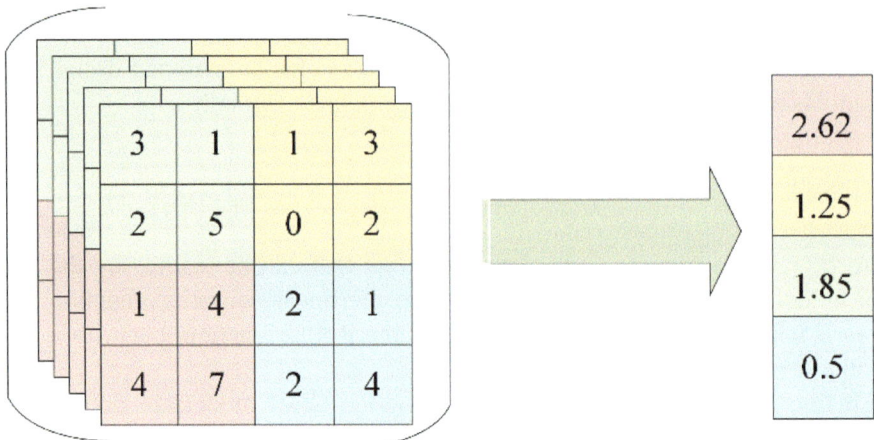

Figure 6.8 Global average pooling.

advantage. The spatial information is summed up in global average pooling, making it more resistant to input translations in space. The global average pooling method of structural regularization requires the use of concept (category) confidence maps that are explicit in their level of certainty.

Finally, the cross-channel technique is employed as a method of local normalization to produce summations from adjacent maps at identical locations. As a result, the feature maps are standardized before being passed on to the next network layer. When using a SoftMax activation function, neurons can only be activated between 0 and 1. Here is how we characterize the SoftMax activation function:

$$\text{SoftMax}(V_f) = \frac{e^{v_f}}{\sum_{g=1}^{w} v_g}, \qquad (6.2)$$

where e^{v_f} represents the exponential function applied to every term v_f within the input vector V, $\sum_{g=1}^{w} v_g$ designates the normalization element, and w stands for the total number of classes.

6.4 Results and Discussion

In this section, the proposed model EfficientNetB3-DTL performance has been described. It is described as follows.

On Kaggle with the keras tensorflow library, the proposed EffientNetB3-DTL model is implemented using GPU P100 accelerator. The investigation is conducted utilizing a laptop equipped with an Intel Core i5 7200U CPU, a dedicated graphics processing unit featuring a Geforce 940 MX, and 8 GB of random-access memory.

The DR dataset is employed to assess the performance of EffientNetB3-DTL model throughout testing and training. The dataset has H_dr of 1000 images, Mi_dr of 370 images, M_dr of 900 images, P_dr 290 images, and S_dr of 190 images. In the experiment, the data is trimmed to 190 images in each class before applying to the EffientNetB3-DTL to give balanced performance for each class. 80% of the data is utilized to train the model, while 20% is used for testing and are illustrated in Figure 6.9.

Sensitivity, specificity, accuracy, precision, recall, and F1 score are employed for evaluating the performance of the proposed EffientNetB3-DTL model. These performance parameters are defined as follows [27–30]:

$$\text{Accuracy} = \frac{\text{TrPos} + \text{TrNeg}}{\text{TrPos} + \text{TrNeg} + \text{FlsPos} + \text{FlsNeg}} \qquad (6.3)$$

$$\text{Sensitivity} = \frac{\text{TrPos}}{\text{TrPos} + \text{FlsNeg}} \qquad (6.4)$$

$$\text{Specificity} = \frac{\text{TrNeg}}{\text{TrNeg} + \text{FlsPos}} \qquad (6.5)$$

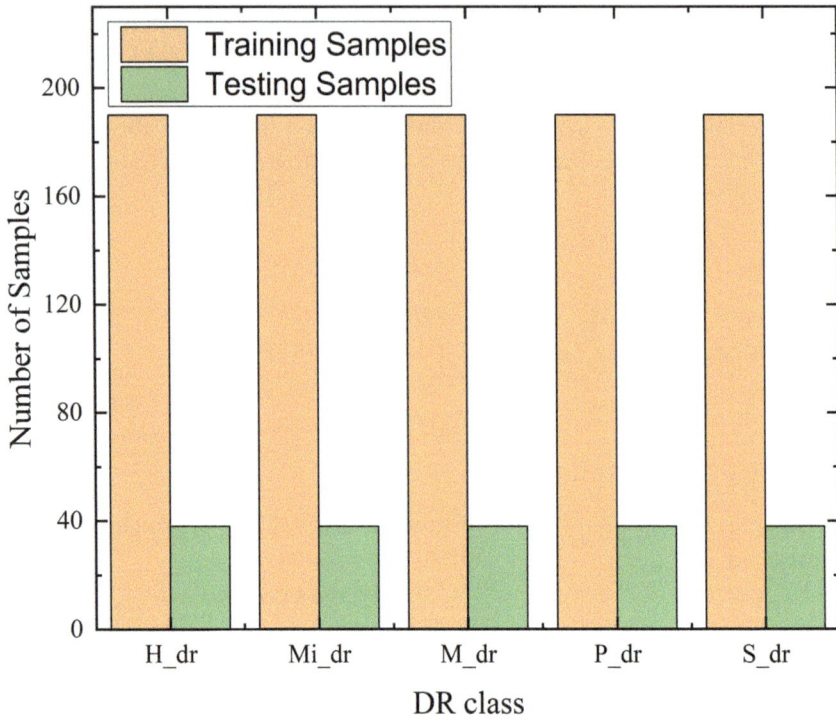

Figure 6.9 Balanced samples for training and testing of the EffientNetB3-DTL model.

$$\text{Precision} = \frac{\text{TrPos}}{\text{TrPos} + \text{FlsPos}} \quad (6.6)$$

$$\text{Recall} = \frac{\text{TrPos}}{\text{TrPos} + \text{TrNeg}} \quad (6.7)$$

$$\text{F1Score} = 2\left[\frac{(\text{recall} \times \text{precision})}{(\text{recall} + \text{precision})}\right] \quad (6.8)$$

The Adam optimizer is renowned for its widespread adoption and usage in numerous applications. Hence to evaluate the proposed model, the Adam optimizer is employed during the training phase and the hyperparameters during experimentation are given in Table 6.1.

The training and testing accuracies of the EffientNetB3-DTL model is illustrated in Figure 6.10. For training model for 20 epochs, the entire network takes training time of 20 minutes 45 seconds. The results in Figure 6.10 show that EffientNetB3-DTL model had higher training accuracy of 94.2% and testing accuracy 84.6%.

6.4 Results and Discussion 139

Table 6.1 Hyperparameter setting.

Parameter	Training	Testing
Mini-batch size	20	20
Learning rate	0.001	0.001
Epochs	20	20

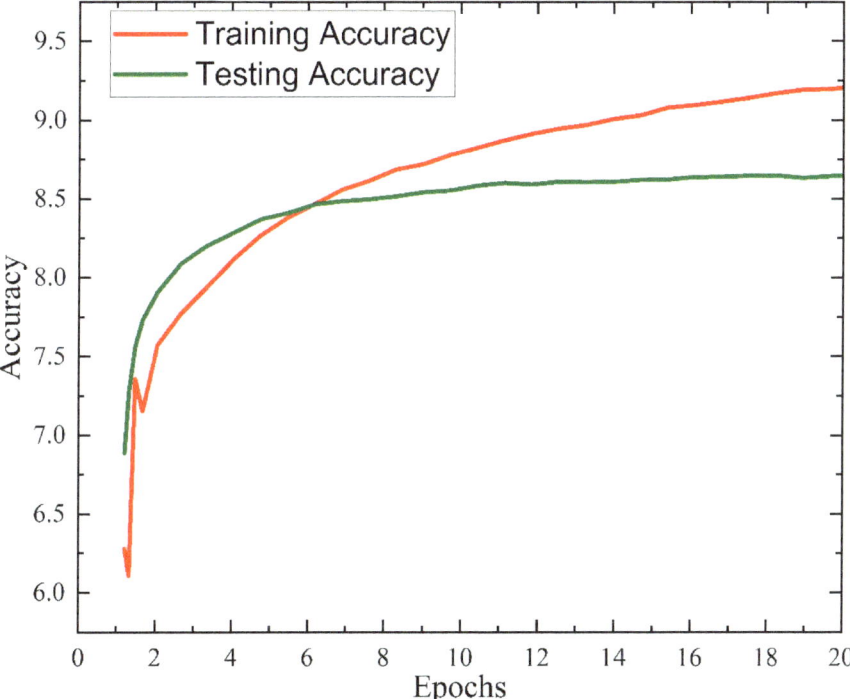

Figure 6.10 Training and testing accuracies of the proposed EffientNetB3-DTL.

The training and testing losses of the EffientNetB3-DTL model is illustrated in Figure 6.11. The results show that EffientNetB3-DTL model lower training loss of 0.4% and testing loss of 0.5%.

Table 6.2 illustrates the simulation parameters for training and verifying the proposed model. The proposed model performs well for H_dr, Mi_dr, and M_dr due to the quality of the images, whereas for P_dr and S_dr, it provides low performance. In addition, the sensitivity and specificity of the model are 84.93%, and 84.89%, respectively.

Finally, the sample test images with the corresponding accuracies are illustrated in Figure 6.12. On DR datasets, EfficientNetB3-DTL classification

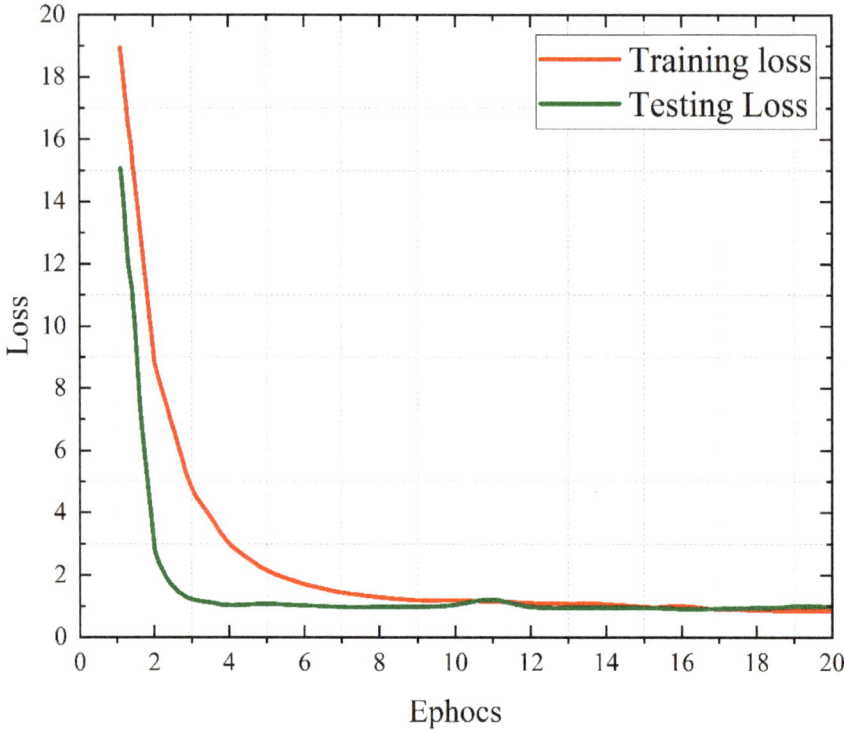

Figure 6.11 Training and testing loss of the proposed EffientNetB3-DTL.

Table 6.2 Performance parameters.

Parameter	H_dr	Mi_dr	M_dr	P_dr	S_dr
Precision	0.99	0.63	0.71	0.61	0.89
Recall	1.00	0.67	0.82	0.50	0.53
F1-score	0.98	0.64	0.76	0.53	0.67
Support	146	24	79	30	15

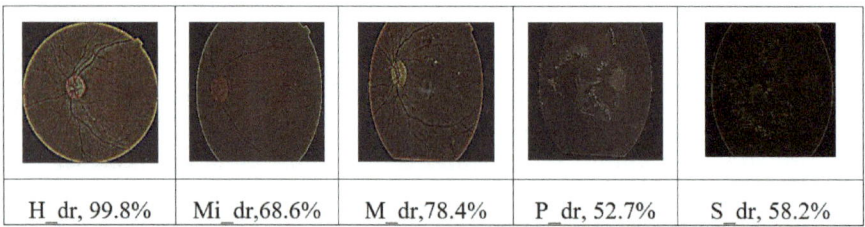

Figure 6.12 The prediction accuracies of the sample images of each class.

accuracy outperforms the best models by around 5%, which was a substantial improvement. This study employed EfficientNetB3 to classify DR images.

6.5 Conclusion

In this chapter, a modified structure of EfficientNetB3-DTL model was presented for diabetic retinopathy classification. The proposed modified EfficientNetB3-DTL model is implemented with MBConv layers with IRB blocks, linear activation function, and GAP layers. The experimental results illustrate that the EfficientNetB3-DTL model had higher accuracy of 84.6%, higher sensitivity of 84.93%, higher specificity of 84.89%, higher precision of 83.24%, and higher F1 score of 83.94%. In addition, the proposed model yielded lower training time of 20 minutes 45 seconds, higher training accuracy of 94.2%, and testing accuracy 84.6% as well as lower training loss of 0.4% and testing loss of 0.5%. In the future, we would like to test the approach by collecting more data samples. Further, improve the method by integrating state-of-the-art deep transfer learning networks

Acknowledgments

We thank SR University, Warangal, India and KITS Warangal, India for supporting us during this work.

References

[1] Harper CA, Keeffe JE. Diabetic retinopathy management guidelines. Expet Rev Ophthalmol 2012;7(5):417–39.
[2] Taylor R, Batey D. Handbook of retinal screening in diabetes:diagnosis and management. second ed. John Wiley & Sons, Ltd Wiley-Blackwell; 2012.
[3] A. Tayal, J. Gupta, A. Solanki, K. Bisht, A. Nayyar, and M. Masud, "Dl cnn-based approach with image processing techniques for the diagnosis of retinal diseases," Multimedia Systems, pp. 1–22, 2021.
[4] Li, Feng, et al. "Automatic detection of diabetic retinopathy in retinal fundus photographs based on deep learning algorithm." Translational vision science & technology 8.6 (2019).
[5] K. Oh, H. M. Kang, D. Leem, H. Lee, K. Y. Seo, and S. Yoon, "Early detection of diabetic retinopathy based on deep learning and

ultra-wide:eld fundus images," Scienti4c Reports, vol. 11, no. 1, pp. 1897–1899, 2021.
[6] Samanta, Abhishek, et al. "Automated detection of diabetic retinopathy using convolutional neural networks on a small dataset." Pattern Recognition Letters 135 (2020): 293-298.
[7] M. Rehman, S. H. Khan, Z. Abbas, and S. D. Rizvi, "Classification of diabetic retinopathy images based on customized cnn architecture," in Proceedings of the 2019 Amity International Conference on Arti4cial Intelligence (AICAI), pp. 244–248, IEEE, Dubai, UAE, February 2019.
[8] Khan, Sharzil Haris, Zeeshan Abbas, and SM Danish Rizvi. "Classification of diabetic retinopathy images based on customised CNN architecture." 2019 Amity International Conference on Artificial Intelligence (AICAI). IEEE, 2019.
[9] L. Qiao, Y. Zhu, and H. Zhou, "Diabetic retinopathy detection using prognosis of microaneurysm and early diagnosis system for non- proliferative diabetic retinopathy based on deep learning algorithms," IEEE Access, vol. 8, Article ID 104292, 2020.
[10] Wang, Xiaoliang, et al. "Diabetic retinopathy stage classification using convolutional neural networks." 2018 IEEE International Conference on Information Reuse and Integration (IRI). IEEE, 2018.
[11] M. Imran, M. Imran, A. Ullah, M. Arif, and R. Noor, "A uni:ed technique for entropy enhancement based diabetic retinopathy detection using hybrid neural network," Computers in Biology and Medicine, vol. 145, Article ID 105424, 2022.
[12] Gargeya, Rishab, and Theodore Leng. "Automated identification of diabetic retinopathy using deep learning." Ophthalmology 124.7 (2017): 962-969.
[13] K. Shankar, Y. Zhang, Y. Liu, L. Wu, and C.-H. Chen, "Hyperparameter tuning deep learning for diabetic retinopathy fundus image classi:cation," IEEE Access, vol. 8, Article ID 118164, 2020.
[14] Pratt, Harry, et al. "Convolutional neural networks for diabetic retinopathy." Procedia computer science 90 (2016): 200-205.
[15] X. Zeng, H. Chen, Y. Luo, and W. Ye, "Automated diabetic retinopathy detection based on binocular siamese-like convolutional neural network," IEEE Access, vol. 7, Article ID 30744, 2019.
[16] Xu, Kele, Dawei Feng, and Haibo Mi. "Deep convolutional neural network-based early automated detection of diabetic retinopathy using fundus image." Molecules 22.12 (2017).

[17] A. J. Reddy, A. Dang, A. A. Dao, G. Arakji, J. Cherian, and H. Brahmbhatt, "A substantive narrative review on the usage of lidocaine in cataract surgery," Cureus, vol. 13, no. 10, Article ID 19138, 2021.

[18] Bellemo, Valentina, et al. "Artificial intelligence using deep learning to screen for referable and vision-threatening diabetic retinopathy in Africa: a clinical validation study." The Lancet Digital Health 1.1 (2019).

[19] https://www.kaggle.com/datasets/sachinkumar413/diabetic-retinopathy-dataset

[20] Prasad, C. R., Kollem, S., Samala, S., Rao, P. R., Yalabaka, S., & Chakradhar, A. (2022, December). Devanagari Script Digit Classification using modified AlexNet with Transfer Learning. In 2022 International Conference on Smart Generation Computing, Communication and Networking (SMART GENCON) (pp. 1-4). IEEE.

[21] Krizhevsky, A., Sutskever, I., & Hinton, G. E. (2017). Imagenet classification with deep convolutional neural networks. Communications of the ACM, 60(6), 84-90.

[22] Alhichri, H., Alswayed, A. S., Bazi, Y., Ammour, N., & Alajlan, N. A. (2021). Classification of remote sensing images using EfficientNet-B3 CNN model with attention. IEEE access, 9, 14078-14094.

[23] Tan, M., & Le, Q. (2019, May). Efficientnet: Rethinking model scaling for convolutional neural networks. In International conference on machine learning (pp. 6105-6114). PMLR.

[24] Sandler, M., Howard, A., Zhu, M., Zhmoginov, A., & Chen, L. C. (2018). Mobilenetv2: Inverted residuals and linear bottlenecks. In Proceedings of the IEEE conference on computer vision and pattern recognition (pp. 4510-4520).

[25] He, K., Zhang, X., Ren, S., & Sun, J. (2016). Deep residual learning for image recognition. In Proceedings of the IEEE conference on computer vision and pattern recognition (pp. 770-778).

[26] A. Kızrak. (Jan. 7, 2020). Comparison of Activation Functions for Deep Neural Networks. Accessed: May. 25, 2023. [Online]. Available: https://towardsdatascience.com/comparison-of-activation-functions-for-deep-neural-networks-706ac4284c8a.

[27] R. Janapati, U. Desai, S. A. Kulkarni, and S. Tayal, Human-Machine Interface Technology Advancements and Applications. CRC Press, 2023.

[28] Ankit Vijayvargiya, Bharat Singh, Rajesh Kumar, Usha Desai, Jude Hemanth, "Hybrid Deep Learning Approaches for sEMG Signal-Based

Lower Limb Activity Recognition", Mathematical Problems in Engineering, vol. 2022, Article ID 3321810, 12 pages, 2022.

[29] Prasad, C. R., Varshamrutha, K., Sindhuja, B., Ushasree, R., Nikhil, P., & Chakradhar, A. (2024, May). MRI Based Brain Tumor classification Using a Fine-Tuned EfficientNetB3 Transfer Learning Model. In 2024 5th International Conference for Emerging Technology (INCET) (pp. 1-5).

[30] P. M. Rajasree, A. Jatti, D. Santosh, U. Desai and V. D. Krishnappa, "Breast Masses Detection and Segmentation in Full-Field Digital Mammograms using Unified Convolution Neural Network," 2022 44th Annual International Conference of the IEEE Engineering in Medicine & Biology Society (EMBC), Glasgow, Scotland, United Kingdom, 2022, pp. 1002-1007.

7

Prediction and Diagnosis of Glaucoma in Fundus Images through Optic Cup and Optic Disk Segmentation

M. Ponnibala[1], Usha Desai[2], Biswaranjan Acharya[3], Vassilis C. Gerogiannis[4], and Andreas Kanavos[5]

[1]Department of Biomedical Engineering,
Velalar College of Engineering and Technology, India
[2]Department of Electronics and Communication Engineering, S.E.A College of Engineering & Technology, Bengaluru, India
[3]Department of Computer Engineering-AI and Big Data Analytics, Marwadi University, India
[4]Department of Digital Systems, University of Thessaly, Greece
[5]Department of Informatics, Ionian University, Greece
E-mail: ponnibala@velalarengg.ac.in; dr.ushadesai@seaedu.ac.in; biswaranjan.acharya@marwadieducation.edu.in; vgerogian@uth.gr; akanavos@ionio.gr

Abstract

Glaucoma, a leading cause of blindness, is often detected early through structural changes in the optic nerve head of the retina. This research proposes a method for segmenting the optic disc and optic cup and extracting retinal features from fundus images. The segmentation is achieved using active contour, level set, and K-means clustering-based techniques. Texture-based feature extraction is then performed on the segmented images to identify the severity level of glaucoma. The cup to disc ratio (CDR) is calculated from the segmented fundus image, and additional features, such as gray-level co-occurrence matrix (GLCM), are utilized for glaucoma detection. Supervised classifiers, namely multi-class support vector machine (SVM) and extreme learning machine (ELM), are employed for glaucoma stage classification. In addition to clinical features, statistically significant texture features are incorporated for classification. The proposed automated diagnostic system achieved an average classification accuracy of 96.66%, with 100% sensitivity and specificity attained using the ELM classifier with texture features. This system holds promise for reducing the workload of medical professionals during mass screening for glaucoma stages.

Keywords: Glaucoma, optic disc, optic cup, feature extraction, K-means clustering, cup to disc ratio (CDR)..

7.1 Introduction

Glaucoma poses a significant concern and challenge in current healthcare, being the second leading cause of blindness with a prevalence of 2.4% for all ages and 4.7% for individuals above 75 years [21]. This progressive optic neuropathy, primarily caused by increased intraocular pressure (IOP) due to impaired eye drainage, leads to irreversible vision loss. Early detection and monitoring of glaucoma are crucial to prevent severe progression and preserve vision. Ophthalmologists regularly examine retinal images obtained using a color fundus camera to assess glaucoma progression.

Glaucomatous damage affects the optic nerve head, leading to the loss of retinal ganglion cells and subsequent vision loss and permanent blindness. As glaucoma often remains asymptomatic until advanced stages, it has earned the moniker "silent thief of sight." Early detection through regular supervision and subsequent medical intervention by ophthalmologists can effectively slow down the disease's progression. Geometric parameters of the optic nerve head (ONH), such as the cup to disc ratio (CDR), play a significant role in diagnosing and measuring glaucoma progression.

Figure 7.1 displays examples of normal and abnormal fundus images in glaucoma. The optic disc, also known as the optic nerve head, serves as the entry point for major blood vessels to the retina. Observing structural changes in the optic disc guides ophthalmologists in assessing the degree of optic nerve damage in glaucoma patients.

Color fundus imaging (CFI), a non-invasive modality, has become a preferred screening tool for assessing glaucoma and other eye conditions, such as diabetic retinopathy and macular edema, on a large scale. This imaging technique allows for efficient acquisition of fundus images and can be

7.1 Introduction   147

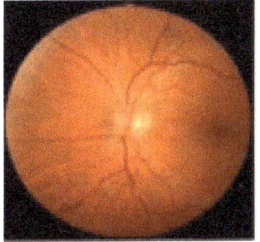
(a) Normal fundus

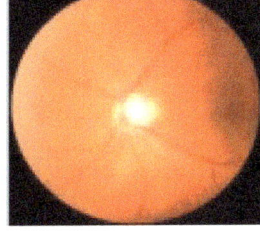
(b) Abnormal fundus

Figure 7.1   Retinal fundus.

integrated into automated systems to assist in identifying affected glaucoma images, reducing the workload of ophthalmologists during examination [5].

Figure 7.2 shows the region of the optic disc and optic cup, as well as the neuroretinal rim located at the back of the eye. The optic nerve head, also referred to as the optic disc, serves as the entry point for major blood vessels into the retina. The color of a normal optic disc is typically orange to pink. However, a pale disc, appearing with a color ranging from pale pink or orange to white, indicates a diseased eye condition.

The physiological cup, a normal slight depression in the optic disc, can be observed by ophthalmologists during examinations. Changes in the size and shape of the cup, specifically enlargement relative to the optic disc, are indicative of glaucoma progression. Parameters such as area, optic disc diameter, optic cup diameter, rim area, and mean cup depth are commonly used to quantify and monitor the progression of glaucoma.

By effectively segmenting and analyzing the optic disc and optic cup regions in fundus images, ophthalmologists can gain valuable insights into the severity of glaucoma and its progression, aiding in early detection and timely intervention to preserve vision.

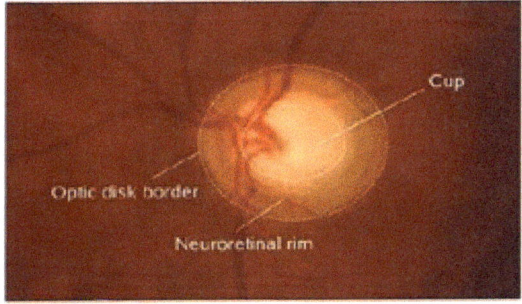

Figure 7.2   Optic disc and optic cup region.

In this study, our primary contribution lies in the development of an automated diagnostic system for the prediction and diagnosis of glaucoma using fundus images. We propose a novel approach that involves segmenting the optic disc and optic cup regions with high accuracy and efficiency using active contour, level set, and K-means clustering-based techniques. Subsequently, we employ texture-based feature extraction methods, including cup to disc ratio (CDR) and gray level co-occurrence matrix (GLCM), to detect glaucoma severity.

The extracted retinal features are then utilized as inputs to two supervised classifiers, namely multi-class support vector machine (SVM) and extreme learning machine (ELM), for accurate glaucoma stage classification. Through extensive experimentation and analysis, we demonstrate that the automated system achieves a remarkable average classification accuracy of 96.66%, with 100% sensitivity and specificity when using the ELM classifier with texture features. This significant outcome suggests the potential of our automated diagnostic system to assist medical professionals in mass screening for glaucoma stages, ultimately reducing their workload and enhancing the efficiency of glaucoma diagnosis.

By presenting this comprehensive and promising approach, we aim to contribute to the field of computer-assisted glaucoma diagnosis, where early detection and timely intervention are critical in preserving patients' vision and mitigating the progression of this debilitating eye disease. Our research endeavors to bridge the gap between advanced computer vision techniques and ophthalmology, bringing us closer to a cost-effective and accessible solution for glaucoma detection and management, thereby improving the quality of eye care and making a positive impact on public health.

With the successful implementation of our automated diagnostic system, we anticipate its integration into clinical practices and large-scale screening programs, enabling early identification and intervention for individuals at risk of glaucoma. This research not only advances the field of computer-aided medical diagnosis but also contributes to the broader goal of reducing the global burden of glaucoma-related blindness.

7.2 Related Work

In this section, we provide an overview of the research efforts related to the segmentation of the optic disc and optic cup in glaucoma diagnosis. We categorize the existing works based on the segmentation techniques employed, highlighting their main contributions and achievements.

7.2.1 Color-based segmentation

The authors in [26] proposed an automatic localization of the optic cup using a combination of the level set and convex hull methods. Their approach included a multimodality fusion technique to estimate the cup boundary and eliminate interference from blood vessels. The method achieved an impressive accuracy of 97.2% on a test dataset of 71 manually segmented images.

In addition, an automated optic disc parameterization technique based on segmented optic disc and cup regions is developed in [9]. Their multidimensional feature space segmentation approach showed robustness against variations around the optic disc region, and a multi-stage strategy was employed for reliable vessel bend identification.

7.2.2 Model-based segmentation

A model-based method for the segmentation of the optic disc and optic cup is proposed in [24]. Their approach combined knowledge-based circular Hough transform with an optimal channel selection, achieving a mean absolute CDR error of 0.10. Similarly, a super pixel-based segmentation method for screening glaucoma is introduced in [4]. They utilized center surround statistics to classify super pixels into disc and non-disc regions and identify the optic cup location.

7.2.3 Texture-based segmentation

The authors in [11] proposed an effective segmentation of the optic disc and optic cup using the fuzzy C-means technique. The method employed the extracted green channel and incorporated dilation and erosion to remove the vernacular region, achieving accurate CDR measurements.

The gradient vector flow-based active contour model algorithm for optic disc segmentation was used in [14]. Their approach involved using nine popular active shape model algorithms to derive the initial OD contour from the circular Hough transform.

A methodology for accurate center position location and optic disc retinal region segmentation using digital fundus images is developed in [13]. Their approach utilized iterative opening–closing morphological operations and automatic thresholding procedures to obtain a reduced region of interest, enabling the extraction of the optic disc boundary [20].

Furthermore, the authors in [15] proposed an optic disc region segmentation method using active contour, incorporating image information from multiple channels to enhance robustness against variations in and around the optic disc region. Additionally, they employed structural and gray-level properties of the cup for precise glaucoma assessment. Also, the whole process of glaucoma diagnosis can be optimized using nature-inspired different algorithms along with the feature extraction and classification process. Recently, the authors in [22] implemented a hybrid method combining group search optimization and particle search optimization to optimize this diagnosis problem. Diabetic retinopathy detection and classification using deep learning can be found in [16] where the authors discussed two deep-learning-based hybrid classifiers for classification.

It is also important to note the work [17], which addresses the challenges of differentiating the optic disc from the peri papillary atrophy (PPA) region. Their contribution provides valuable insights into managing the PPA region, which is often mistaken as part of the optic disc due to its similar appearance and semi-circular shape, resulting in potential misinterpretations.

Based on the comprehensive study of these published works, it is evident that color-based segmentation methods are well-suited for early detection of the optic disc and optic cup regions. Glaucoma diagnosis is often based on shape-based, pixel-based, and texture-based features. However, the existing methods have challenges in differentiating the optic disc from the peri papillary atrophy (PPA) region, which may lead to misinterpretations. The proposed segmentation technique in this chapter employs active contour, level set, and K-means clustering to accurately delineate the optic disc and cup regions, effectively addressing these challenges (Figure 7.3).

By presenting this comprehensive overview, we establish the context for our novel approach, which incorporates a combination of segmentation techniques and texture-based feature extraction to develop an automated diagnostic system for glaucoma prediction and diagnosis.

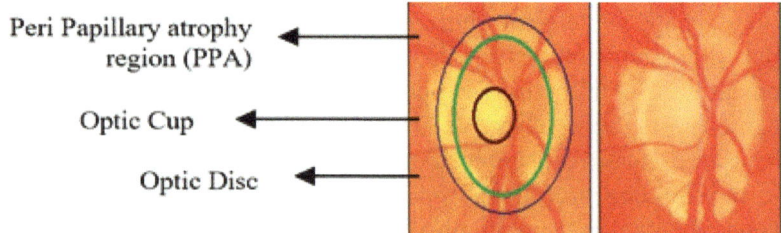

Figure 7.3 PPA and optic disc, optic cup region.

7.3 Material and Methods

The main objective of this work is to extract the optic disc and optic cup for the assessment of glaucoma using features extracted from preprocessed retinal images. The image obtained from the database undergoes preprocessing, followed by segmentation of the optic disc, optic cup region, and cup to disc ratio (CDR) measurement. Three segmentation techniques are proposed: gradient-based active contour method, level set algorithm, and K-means clustering-based segmentation technique. These techniques are employed to differentiate between the optic disc, optic cup region, and neuroretinal rim. The resulting images are then utilized for feature extraction, including CDR and texture properties. Selected features are used in the diagnosis of glaucoma.

7.3.1 Retinal image acquisition

In this study, retinal images with and without abnormalities are obtained from the DRIONS-DB (Digital Retinal Image for Optic Nerve Segmentation Database) [2]. The DRIONS database contains 110 images acquired by a color analogical fundus camera in RGB format with a resolution of 600 × 400 and 8 bits per pixel.

7.3.2 Proposed methodology

The proposed methodology comprises four stages: region of interest (ROI) extraction, segmentation, feature extraction, and classification. The block diagram of the proposed methodology is shown in Figure 7.4.

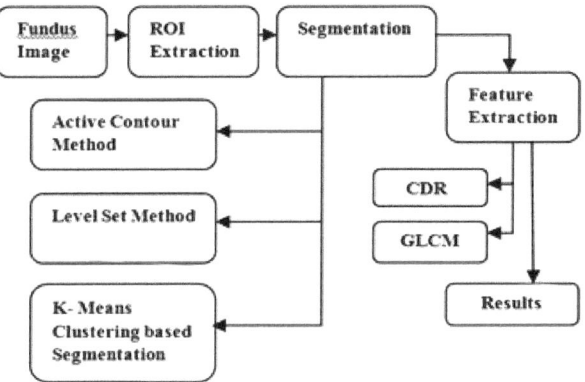

Figure 7.4 Block diagram of the proposed method.

The three segmentation techniques – gradient-based active contour method, level set algorithm, and K-means clustering – are used to extract the optic disc and optic cup. The ROI is initially selected around the brightest point in the pre processed grayscale image, forming a square of size 360 × 360 pixels with the brightest pixel at the center. This initial ROI covers the entire optic disc, optic cup region, and a small portion of other regions of the image.

7.3.3 Active contour segmentation for optic disc localization

The active contour method provides an effective way to segment the optic disc by detecting its boundaries through curve evolution. The gradient-based active contour model is utilized for disc boundary detection. The algorithm aims to minimize the energy function shown in eqn (7.1):

$$E_{\text{snake}} = E_{\text{internal}} + E_{\text{external}} + E_{\text{constraint}}. \tag{7.1}$$

The initial contour points are manually localized to identify the boundary of the optic disc region. The optic disc regions are segmented from the pre processed grayscale image, as shown in Figure 7.5. Additionally, the ChanVese (CV) model, another active contour model, is used for refining the OD region segmentation by integrating information from multiple image feature channels [1, 14, 18].

7.3.4 Level set algorithm for optic cup localization

The level set method is used for contour evolution in the segmentation process. The function $\phi(i, j, t)$ (level set function) is defined, where (i, j) are coordinates in the image plane, and t represents time. At any given time, the level set function simultaneously defines an edge contour and a segmentation of the image. The edge contour is the zero-level set $(i, j) \mid \phi(i, j, t) = 0$, and

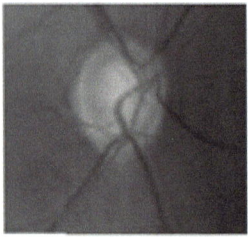 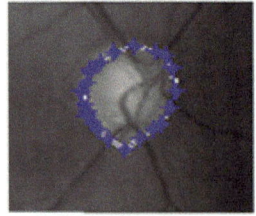 

(a) Original fundus   (b) Initialization of contour   (c) Segmented optic disc

Figure 7.5 Results of active contour technique.

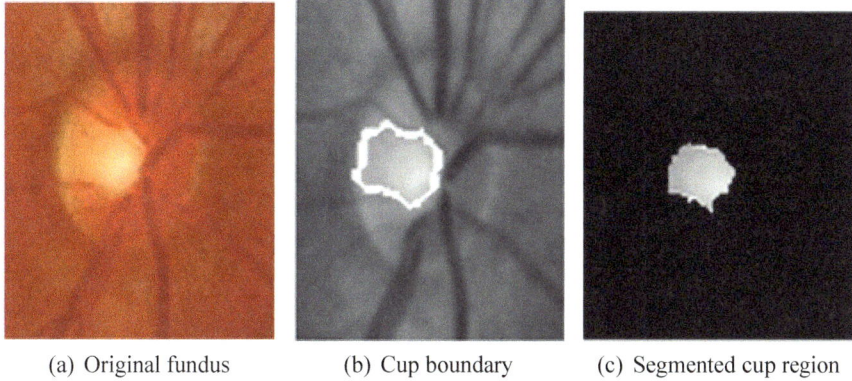

(a) Original fundus   (b) Cup boundary   (c) Segmented cup region

Figure 7.6 Results of level set technique.

the segmentation is defined by the two regions $\phi \geq 0$ and $\phi < 0$. The level set function evolves according to some partial differential equation, allowing it to handle topological changes in the edge contour that might be challenging to manage with a direct contour evolution model [13]. Figure 7.6 illustrates the results of optic cup segmentation using the level set method.

7.3.5 K-means clustering based segmentation algorithm

The K-means clustering algorithm is employed to cluster the fundus image into different clusters based on their properties. The proposed work uses the K-means algorithm to cluster the fundus image into three clusters of red, blue, and green. The algorithm aims to partition "n" observations into "k" clusters, with each observation belonging to the cluster with the nearest mean [3, 6, 7, 10, 12]. After the clustering process, the segmentation of the optic disc and optic cup is performed. The thresholding technique is applied to compare each pixel's gray level with a single threshold image, effectively isolating the objects and converting grayscale images into binary images.

The clustered fundus images are shown in Figure 7.7. Each cluster represents a specific region, such as the optic disc, optic cup, and other areas in the image. The K-means clustering plays a crucial role in distinguishing between the different regions, which facilitates the subsequent segmentation of the optic disc and optic cup.

After the K-means clustering process, the fundus image is divided into distinct clusters representing different regions based on their properties. Among these clusters, the one corresponding to the optic disc and optic cup is identified. However, the information about the exact boundary of these regions is still contained within the grayscale image.

To obtain a binary segmented image that clearly distinguishes the optic disc and optic cup regions, a thresholding technique is applied. This technique involves comparing the grayscale intensity of each pixel in the image with a predefined threshold value. All pixels with intensity values below the threshold are classified as part of one phase, while those with intensity values above the threshold are assigned to the other phase. In this case, the desired regions are the optic disc and optic cup, which have distinct grayscale properties compared to the background and other structures in the image.

By setting the appropriate threshold value, the regions of the optic disc and optic cup are segmented separately, and the other structures in the image are removed from consideration. This process effectively isolates the regions of interest, allowing for further analysis and feature extraction.

Figure 7.8 illustrates the binary segmented disc and cup region, where the optic disc and optic cup are represented as separate white regions against a black background. These segmented regions provide a clear delineation of the optic disc and optic cup, which is essential for subsequent feature extraction and glaucoma classification.

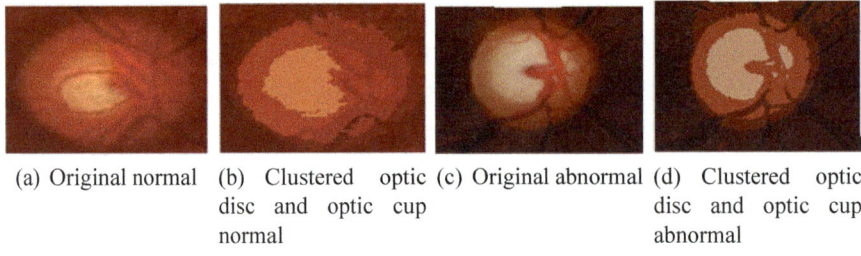

(a) Original normal (b) Clustered optic disc and optic cup normal (c) Original abnormal (d) Clustered optic disc and optic cup abnormal

Figure 7.7 Clustering results.

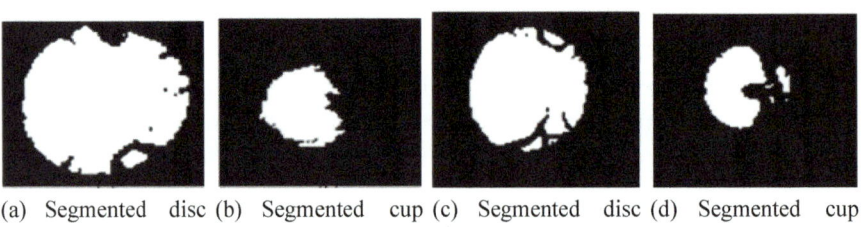

(a) Segmented disc region normal (b) Segmented cup region normal (c) Segmented disc region abnormal (d) Segmented cup region abnormal

Figure 7.8 Segmented optic disc and optic cup.

7.4 Feature Extraction on the Segmented Optic Disk and Cup

Feature extraction plays a crucial role in assessing glaucoma by computing quantitative information from the segmented optic disc and optic cup. These extracted features are essential for classifying objects based on predetermined criteria, including size, color, and texture. The ultimate goal is to identify the progression of glaucoma and categorize it into different stages: normal stage, moderate stage, and severely affected stage.

7.4.1 Cup to disc ratio (CDR)

One of the key clinical features used for assessing glaucoma severity is the cup to disc ratio (CDR). The optic disc is the white, cup-like area located at the center of the optic nerve head. The CDR is calculated by determining the ratio of the vertical cup diameter to the vertical disc diameter. A normal CDR typically falls within the range of 0.1 – 0.3, while a higher CDR above 0.6 indicates a higher risk of glaucoma [3, 11].

In the proposed work, the segmented optic disc and optic cup regions are used to calculate the CDR. The vertical cup diameter and vertical disc diameter are measured from the corresponding boundary layers of the disc and cup. Using the obtained measurements, the CDR is computed using the following formula:

$$\mathrm{CDR} = \frac{\text{Vertical Cup Diameter}}{\text{Vertical Disc Diameter}}. \tag{7.2}$$

Figure 7.9 shows the boundary layer extraction of the optic disc and optic cup, which is a crucial step in calculating the cup to disc ratio (CDR) for the assessment of glaucoma severity. In Figure 7.9, the two subfigures represent the boundary layers of the optic cup and optic disc, respectively. The boundary layer extraction process involves precisely delineating the edges of the optic cup and optic disc regions from the segmented images obtained after the thresholding and segmentation process.

By accurately extracting these boundary layers, the vertical cup diameter and vertical disc diameter can be measured, and the CDR can be computed using the formula mentioned earlier. The CDR serves as an essential quantitative parameter for diagnosing and monitoring glaucoma, as it provides valuable insights into the progression of the disease.

156 *Prediction and Diagnosis of Glaucoma in Fundus Images*

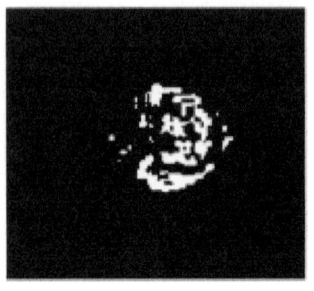 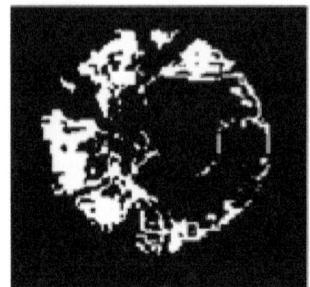

(a) Cup boundary (b) Disc boundary

Figure 7.9 CDR results.

The boundary layer extraction shown in Figure 7.9 is part of the feature extraction process, which plays a vital role in glaucoma assessment and aids clinicians in making informed decisions about the patient's condition.

Table 7.1 presents the standard baseline data for glaucoma, providing the cup to disc ratio (CDR) ranges that correspond to different stages of the disease.

The cup to disc ratio (CDR) is a critical clinical parameter used to assess the severity of glaucoma. It is calculated using the boundary layers obtained from the segmented optic disc and optic cup regions, as shown in Figure 7.9. The boundary layers are extracted from the segmented images through an accurate and precise process.

The feature extraction stage involves measuring the vertical cup diameter (VCD) and vertical disc diameter (VDD) from the respective boundary layers. Once the CDR values are obtained, they are compared with the ranges provided in Table 7.1 to diagnose the stage of glaucoma. By categorizing the CDR results into these stages, clinicians can effectively classify and diagnose glaucoma cases, allowing for appropriate treatment and monitoring.

Table 7.2 displays the computed CDR values for a set of 40 images using the RIM-ONE database, along with the corresponding stages of glaucoma diagnosis.

Table 7.1 Standard baseline data for glaucoma.

Stages	CDR range
Normal	0.1 – 0.3
Mild glaucoma	0.3 – 0.6
Severe glaucoma	> 0.6

7.4 Feature Extraction on the Segmented Optic Disk and Cup

Table 7.2 Results of cup to disc ratio.

No.	Image	Cup height	Disc height	CDR	Stages of glaucoma
1	Image_001	24	81	0.2963	Normal eye
2	Image_002	32	77	0.4156	Mild stage
3	Image_003	32	85	0.3765	Mild stage
4	Image_004	64	99	0.6465	Abnormal eye
5	Image_005	36	95	0.3789	Mild stage
6	Image_006	98	108	0.9074	Abnormal
7	Image_007	52	102	0.5098	Mild stage
8	Image_008	82	128	0.6406	Abnormal eye
9	Image_009	65	108	0.6019	Abnormal eye
10	Image_010	45	92	0.4891	Mild stage
11	Image_011	45	92	0.4891	Mild stage
12	Image_012	25	79	0.3164	Normal eye
13	Image_013	27	79	0.3418	Normal eye
14	Image_025	48	89	0.5393	Mild stage
15	Image_065	38	72	0.5278	Mild stage
16	Image_067	79	106	0.7452	Abnormal eye
17	Image_068	92	113	0.814	Abnormal eye
18	Image_079	23	29	0.7931	Abnormal eye
19	Image_069	74	107	0.6915	Abnormal eye
20	Image_070	72	105	0.6857	Abnormal eye
21	Image_065	30	109	0.2752	Normal eye
22	Image_069	48	111	0.4324	Mild stage
23	Image_091	85	130	0.6538	Abnormal eye
24	Image_092	45	92	0.4891	Mild stage
25	Image_094	41	108	0.3796	Mild stage
26	Image_097	91	111	0.8198	Abnormal eye
27	Image_099	69	94	0.734	Abnormal eye
28	Image_102	68	86	0.7907	Abnormal eye
29	Image_027	53	92	0.5760	Mild stage
30	Image_028	55	82	0.6707	Abnormal eye
31	Image_035	45	92	0.4892	Mild stage
32	Image_041	22	70	0.3142	Normal eye
33	Image_058	51	99	0.5152	Mild stage
34	Image_034	42	78	0.5385	Mild stage
35	Image_036	40	120	0.3333	Normal eye
36	Image_040	59	81	0.7284	Abnormal eye
37	Image_042	65	90	0.7222	Abnormal eye
38	Image_056	29	80	0.3625	Mild stage
39	Image_058	45	91	0.4945	Mild stage
40	Image_063	79	110	0.7182	Abnormal eye

By categorizing the CDR values into these stages, ophthalmologists and clinicians can effectively classify and diagnose glaucoma cases, allowing for appropriate treatment and monitoring. The feature extraction process, which involves computing the CDR values, provides valuable information for the assessment of glaucoma severity. This information assists in accurate diagnosis and treatment planning, contributing to better patient care and management.

7.4.2 Texture-based feature extraction

Gray-level co-occurrence matrix (GLCM) is a statistical method used for examining textures, which considers the spatial relationship of pixels and is employed for extracting second-order statistical features [7, 19]. GLCM describes the frequency of one gray tone appearing in a specified spatial linear relationship with another gray tone within the area of interest. In this work, GLCM features are extracted from the segmented image, resulting in 20 statistical features for texture-based feature extraction. To determine the most relevant features for further classification, those with higher variance values among all the extracted features are selected. The selected features include auto-correlation, cluster shade, sum of square, sum average, and sum variance, as they exhibit significant variance values. Conversely, other features with low variance (approximately equal to zero) are neglected during the subsequent classification process.

Using GLCM, 20 Haralick features were extracted from various categories of images. The average of the GLCM matrices is formulated as the mean GLCM, and from this formulated mean, statistical features are derived, as listed in Table 7.3. The optimal features for classification are then determined by calculating the variance of the first-order statistical function.

The features with higher variance values, which are selected for further classification, are highlighted in Table 7.4. These features, including auto-correlation, cluster shade, sum of square, sum average, and sum variance, serve as input data for testing and training during the classification process. The ranges for these selected features are mentioned in Table 7.4, based on which the classification is performed.

The classification process is based on these selected features, and their ranges aid in accurately categorizing and diagnosing glaucoma. By leveraging these texture-based features, the system can effectively assess glaucoma severity, contributing to improved patient care and management.

Table 7.3 Features selection from GLCM.

Features extracted	Normal	Moderate stage	Severe stage	Mean	Variance
Auto correlation	8.5647	16.8219	23.1219	16.1695	53.29724
Contrast	0.023	0.0218	0.0486	0.031133	0.000229
Correlation	0.9999	1	1	0.999967	3.33e-09
Cluster prominence	−1.2279	−1.2685	−1.2854	−1.2606	0.000873
Dissimilarity	0.023	0.0218	0.0486	0.031133	0.000229
Energy	0.4302	0.3974	0.197	0.341533	0.015936
Entropy	−0.3479	−0.3264	−0.1781	−0.28413	0.008548
Homogeneity	0.9885	0.9891	0.9757	0.984433	5.73e-05
Maximum probability	0.5876	0.5524	0.2614	0.467133	0.032054
Cluster shade	1.0277	1.0794	1.7438	1.283633	0.159483
Sum of square	8.4888	16.7041	22.998	16.06363	52.93687
Sum average	3.7251	6.099	7.2153	5.6798	3.17717
Sum variance	8.7808	26.9115	37.5515	24.4146	21.6142
Sum entropy	0.023	0.0218	0.0486	0.031133	0.000229
Difference variance	0.023	0.0218	0.0486	0.031133	0.000229
Difference entropy	0.1096	0.1051	0.1944	0.136367	0.002531
Inverse measure of correlation 1	−2.3745	−2.3304	−2.1138	−2.2729	0.019471
Inverse measure of correlation 2	0.9939	0.9949	0.9993	0.996033	8.25e-06
Inverse difference normalized	0.9974	0.9976	0.9946	0.996533	2.81e-06
Inverse difference moment normalized	0.9996	0.9997	0.9993	0.999533	4.33e-08

Table 7.4 Features range.

Selected Features	Range	
	Min	Max
Auto correlation	8.5647	16.1695
Cluster shade	1.0277	1.7438
Sum of square	8.4888	22.998
Sum average	3.7251	7.2153
Sum variance	8.7808	37.5515

7.5 Experimental Analysis

7.5.1 Multi-class support vector machine (SVM) classifier

In the experimental analysis, the input images undergo segmentation, and the clinical feature CDR and texture features are then extracted from the resultant images. These extracted features are utilized as inputs for the multi-class support vector machine (SVM) classifier to identify the classification accuracy and the severity level of the disease.

Support vector machine (SVM), introduced by Vapnik, is a powerful learning method primarily used for binary classification. The underlying principle of SVM is to find a hyperplane that optimally separates the d-dimensional data into its two classes. This method belongs to the category of supervised classification algorithms, where a training set of known objects is used to build a decision function.

In the training set, each object is represented by a feature vector, and it is associated with a corresponding class value. The learning algorithm leverages this information to create a decision function that can be used to classify new, previously unseen data. Initially designed for binary classification, SVM has been extended to handle multi-class scenarios by decomposing them into a series of binary problems.

Two representative ensemble schemes for multi-class SVM are the one-versus-rest (1VR) and one-versus-one (1V1) approaches. Both of these methods decompose the multi-class problem into a predefined set of binary problems. Another approach involves directly addressing the multi-class problem in a single optimization process, combining multiple binary-class optimization problems into one objective function, which allows for simultaneous multi-class classification.

To apply SVM effectively to any classification problem, several user-defined parameters need to be determined. These parameters include the choice of a suitable multi-class approach, selection of an appropriate kernel and its related parameters, determination of the optimal value of the regularization parameter (C), and the choice of an appropriate optimization technique.

Instead of creating multiple binary classifiers to determine class labels, the multi-class SVM method attempts to directly solve the multi-class problem. This is achieved by modifying the binary class objective function and introducing a constraint for every class. The modified objective function enables simultaneous computation of multi-class classification.

$$\min_{W,b,\zeta} \left[\frac{1}{2} \sum_{i=1}^{M} ||w||^2 + C \sum_{i=1}^{k} \sum_{r \neq y_i} \xi_i^r \right], \qquad (7.3)$$

subject to the constraints

$$W_{y_i}.X_i + b_{y_i} \geq W_r.X_i + b_r + 2 - \xi_i^r, \qquad (7.4)$$
$$\xi_i^r \geq 0 \quad \text{for } i = 1, \ldots, k. \qquad (7.5)$$

where $y_i \in 1, \ldots, M$ represents the multi-class labels of the data vector, and $r \in 1, \ldots, M \setminus y_i$ are the multi-class labels excluding y_i.

In the proposed work, multi-class support vector machine (SVM) is employed for the classification task. It offers several advantages, such as fast convergence and a reduction in the search space dimension compared to conventional back-propagation networks. The performance of the SVM classifier is evaluated based on the classification accuracy, sensitivity, and specificity for different feature sets, including CDR alone and a combination of CDR and GLCM features.

7.5.2 Extreme learning machine (ELM) classifier

The extreme learning machine (ELM) is a single hidden layer feedforward neural network, where the input weights are randomly chosen, and the output weights are analytically calculated [8]. In the context of single-hidden layer feedforward neural networks (SLFN), if the activation functions in the hidden layer are infinitely differentiable, the weights of the input layer and the hidden layer biases can be randomly assigned for the ELM network. Once the weights and hidden layer biases are randomly selected, the SLFNs can be treated as a linear system, and the output weights of SLFNs can be analytically determined using a simple generalized inverse operation on the output matrices of the hidden layer. This leads to the development of the ELM learning algorithm, which exhibits a faster learning speed compared to conventional feedforward network learning algorithms like the back-propagation (BP) algorithm, and also demonstrates better generalization performance.

The ELM classifier is a three-step algorithm that does not require a tuning mechanism. It is known for its extremely fast learning speed, making it well-suited for use as a classifier. Originally developed for SLFNs, ELM was later extended to "generalized" SLFNs. ELM not only aims to achieve the smallest training error but also the smallest norms of output weights [23]. When compared to conventional learning methods, ELM can generate the hidden node parameters before the training data is fed to the network. It consists of only one hidden layer, where the input layer to hidden layer weights can be randomly chosen, and the hidden layer to output layer weights can be calculated analytically.

Unlike traditional gradient-based learning algorithms, which only work with differentiable activation functions, ELM can handle all bounded non-constant activation functions [25]. ELM's straightforward approach in reaching solutions sets it apart from traditional gradient-based learning algorithms, which may encounter issues such as getting stuck in local minima and improper learning rates. ELM's learning algorithm is simpler and requires significantly less training time compared to popular algorithms like the back-propagation (BP) algorithm. Additionally, ELM does not require tuning, and its implementation is easier compared to BP and SVM algorithms. Notably, the classification accuracy of ELM is superior to BP in various challenging classification applications.

The results of the performance of both the multi-class SVM and ELM classifiers for the detection of stages of glaucoma on the RIM-ONE database are presented in Table 7.5.

As shown in Table 7.5, the sensitivity, specificity, and accuracy obtained for the SVM classifier with CDR feature alone are 98.3%, 90.9%, and 70% respectively. When a combination of CDR and GLCM features is used, the sensitivity, specificity, and accuracy values are 96%, 99%, and 90% respectively.

For the ELM classifier, the sensitivity, specificity, and accuracy obtained with CDR feature alone are 99.3%, 99.3%, and 93.33% respectively. With a combination of CDR and GLCM features, the sensitivity, specificity, and accuracy values are 100%, 100%, and 96.66%, respectively. The ELM classifier achieves the best overall performance (96.66%) with three-fold cross-validation.

Based on the results, it is evident that the combination of CDR and GLCM features using the ELM classifier yields the best performance for the detection of stages of glaucoma, demonstrating its effectiveness as a powerful classification technique for this medical application.

Table 7.5 Results of sensitivity, specificity and accuracy values of different classifiers.

Features	Sensitivity	Specificity	Accuracy
Multi-class SVM classifier			
CDR	98.3	90.9	70
GLCM + CDR	96	99	90
ELM classifier			
CDR	99.3	99.3	93.33
GLCM + CDR	100	100	96.66

7.5.3 Discussion

In this study, we employed two different classifiers, namely the multi-class support vector machine (SVM) and the extreme learning machine (ELM), to classify and identify the stages of glaucoma based on clinical features and texture-based features extracted from segmented retinal images. The performance of both classifiers was evaluated in terms of sensitivity, specificity, and overall accuracy.

The multi-class SVM classifier is a well-established and widely used method for binary and multi-class classification tasks. In our experiment, the SVM classifier demonstrated good performance with the clinical feature cup-to-disc ratio (CDR) alone, achieving a sensitivity of 98.3% and a specificity of 90.9%. However, when combined with texture-based features extracted using the gray level co-occurrence matrix (GLCM), the classifier's performance improved significantly, with sensitivity and specificity reaching 96% and 99%, respectively. The SVM classifier's strength lies in its ability to find an optimal hyperplane that maximizes the margin between classes, leading to effective separation of data points. Despite its high accuracy with the combined features, the SVM classifier still showed limitations in identifying certain subtle patterns in the data, which may affect its performance in challenging cases of glaucoma diagnosis.

On the other hand, the extreme learning machine (ELM) classifier, a single-hidden-layer feedforward neural network, demonstrated remarkable performance in this study. When using only the CDR feature, the ELM classifier achieved an impressive sensitivity and specificity of 99.3% each, outperforming the SVM classifier in this aspect. When combined with texture-based features, the ELM classifier reached even higher accuracy, with sensitivity, specificity, and overall accuracy reaching 100%, 100%, and 96.66%, respectively. The ELM classifier's success can be attributed to its fast learning speed, ability to handle non-differentiable activation functions, and its straightforward approach in reaching solutions. It also does not require the tuning that is often necessary for other classifiers, making it an attractive choice for medical image classification tasks.

The comparison of the two classifiers indicates that the ELM classifier outperformed the SVM classifier in terms of accuracy and sensitivity. The ELM classifier demonstrated superior performance in correctly identifying glaucoma cases and non-glaucoma cases, and it achieved the highest overall accuracy of 96.66% with the combined features. Additionally, the ELM classifier showed better generalization capabilities and required less time for training compared to the SVM classifier.

Overall, the results suggest that the combination of clinical features (CDR) and texture-based features (GLCM) using the ELM classifier provides the most accurate and reliable method for classifying glaucoma stages in retinal images. The findings of this study have significant implications for the development of computer-aided diagnosis systems for glaucoma, enabling early detection and effective treatment planning, which are critical for preventing vision loss and improving patient outcomes. However, further research and validation on larger datasets and with diverse populations would be beneficial to ensure the robustness and generalizability of the proposed classification approach.

7.6 Conclusions and Future Work

Glaucoma remains a significant cause of blindness worldwide, necessitating the development of accurate and cost-effective techniques for its detection and classification. In this research, we proposed a comprehensive approach for the detection and classification of different stages of glaucoma using retinal images. The proposed method involves the segmentation of the optic disc and optic cup, extraction of shape-based features, and texture-based feature extraction for accurate diagnosis.

The segmentation of the optic disc and optic cup using K-means clustering techniques allowed for the computation of the cup-to-disc ratio (CDR) for each retinal image. The CDR, a structural feature, was found to be effective in detecting glaucoma, with a ratio value exceeding 0.6 indicating a recommendation for further analysis and timely medical intervention to prevent vision deterioration.

Texture-based features were extracted using the gray-level co-occurrence matrix (GLCM) from the segmented optic disc and optic cup regions. These features, in combination with the CDR, proved to be more effective in detecting different stages of glaucoma compared to existing methods.

Two classification algorithms, namely the multi-class support vector machine (SVM) and the extreme learning machine (ELM) classifiers, were utilized to classify the extracted features for glaucoma diagnosis. The ELM classifier demonstrated superior performance, achieving an average accuracy of 96.66% with 100% sensitivity and specificity. The ELM classifier's capability to efficiently classify different stages of glaucoma indicates its potential for clinical applications, aiding ophthalmologists in mass screening of fundus images.

In conclusion, the proposed approach offers a successful grading method for the diagnosis of glaucoma, reducing the workload of clinicians during the mass screening of retinal images. The combination of CDR and GLCM features using the ELM classifier provided the most accurate and reliable classification of glaucoma stages.

For future work, the analysis of new feature detection methods and the use of advanced machine learning classification algorithms can be explored to further improve the severity level classification of glaucoma. Additionally, validation on larger and diverse datasets is necessary to ensure the robustness and generalizability of the proposed approach. The system's integration into clinical practice can significantly contribute to early glaucoma detection and timely intervention, ultimately improving patient outcomes and reducing the global burden of glaucoma-related blindness.

References

[1] Arturo Aquino, Manuel Emilio Gegúndez-Arias, and Diego Marín. Detecting the optic disc boundary in digital fundus images using morphological, edge detection, and feature extraction techniques. *IEEE Transactions on Medical Imaging*, 29(11):1860–1869, 2010.

[2] Enrique J. Carmona, Mariano Rincón, Julián García-Feijoó, and José M. Martínez-de-la-Casa. Identification of the optic nerve head with genetic algorithms. *Artificial Intelligence in Medicine*, 43(3):243–259, 2008.

[3] S. Chandrika and K. Nirmala. Analysis of CDR detection for glaucoma diagnosis. *International Journal of Engineering Research and Application (IJERA)*, 2(4):23–27, 2013.

[4] Jun Cheng, Jiang Liu, Yanwu Xu, Fengshou Yin, Damon Wing Kee Wong, Ngan Meng Tan, Dacheng Tao, Ching Yu Cheng, Tin Aung, and Tien Yin Wong. Superpixel classification based optic disc and optic cup segmentation for glaucoma screening. *IEEE Transactions on Medical Imaging*, 32(6):1019–1032, 2013.

[5] Behdad Dashtbozorg, Ana Maria Mendonça, and Aurélio J. C. Campilho. Optic disc segmentation using the sliding band filter. *Computers in Biology and Medicine*, 56:1–12, 2015.

[6] Nilanjan Dey, Anamitra Bardhan Roy, Achintya Das, and Sheli Sinha Chaudhuri. Optical cup to disc ratio measurement for glaucoma

diagnosis using harris corner. In *3rd IEEE International Conference on Computing, Communication and Networking Technologies (ICCCNT)*, pages 1–5, 2012.

[7] Muhammad Salman Haleem, Liangxiu Han, Jano van Hemert, and Baihua Li. Automatic extraction of retinal features from colour retinal images for glaucoma diagnosis: A review. *Computerized Medical Imaging and Graphics*, 37(7-8):581–596, 2013.

[8] Guang-Bin Huang, Ming-Bin Li, Lei Chen, and Chee Kheong Siew. Incremental extreme learning machine with fully complex hidden nodes. *Neurocomputing*, 71(4-6):576–583, 2008.

[9] Gopal Datt Joshi, Jayanthi Sivaswamy, and S. R. Krishnadas. Optic disk and cup segmentation from monocular color retinal images for glaucoma assessment. *IEEE Transactions on Medical Imaging*, 30(6):1192–1205, 2011.

[10] K Kavitha and M Malathi. Optic disc and optic cup segmentation for glaucoma classification. *International Journal of Advanced Research in Computer Science (IJARCS))*, 2(1):87–90, 2014.

[11] Noor Elaiza Abdul Khalid, Noorhayati Mohamed Noor, and Norharyati Md. Ariff. Fuzzy c-means (fcm) for optic cup and disc segmentation with morphological operation. *Procedia Computer Science*, 42:255–262, 2014.

[12] V. Mahalakshmi and S. Karthikeyan. Clustering based optic disc and optic cup segmentation for glaucoma detection. *International Journal of Innovative Research in Computer and Communication Engineering*, 2(4):3756–3761, 2014.

[13] Diego Marín, Manuel Emilio Gegúndez-Arias, Angel Suero, and José Manuel Bravo. Obtaining optic disc center and pixel region by automatic thresholding methods on morphologically processed fundus images. *Computer Methods and Programs in Biomedicine*, 118(2):173–185, 2015.

[14] M. Caroline Viola Stella Mary, Elijah Blessing Rajsingh, J. Kishore Kumar Jacob, D. Anandhi, Umberto Amato, and S. Easter Selvan. An empirical study on optic disc segmentation using an active contour model. *Biomedical Signal Processing and Control*, 18:19–29, 2015.

[15] Pardha Saradhi Mittapalli and Giri Babu Kande. Segmentation of optic disk and optic cup from digital fundus images for the assessment of glaucoma. *Biomedical Signal Processing and Control*, 24:34–46, 2016.

[16] Cheena Mohanty, Sakuntala Mahapatra, Biswaranjan Acharya, Fotis Kokkoras, Vassilis C. Gerogiannis, Ioannis Karamitsos, and Andreas Kanavos. Using deep learning architectures for detection and classification of diabetic retinopathy. *Sensors*, 23(12):5726, 2023.

[17] Sandra Morales, Valery Naranjo, Jesús Angulo, and Mariano Alcañiz Raya. Automatic detection of optic disc based on PCA and mathematical morphology. *IEEE Transactions on Medical Imaging*, 32(4):786–796, 2013.

[18] Chisako Muramatsu, Toshiaki Nakagawa, Akira Sawada, Yuji Hatanaka, Takeshi Hara, Tetsuya Yamamoto, and Hiroshi Fujita. Automated segmentation of optic disc region on retinal fundus photographs: Comparison of contour modeling and pixel classification methods. *Computer Methods and Programs in Biomedicine*, 101(1):23–32, 2011.

[19] K. Narasimhan and K. Vijayarekha. An efficient automated system for glaucoma detection using fundus image. *Journal of Theoretical and Applied Information Technology*, 33(1):104–110, 2011.

[20] Jan Odstrcilík, Radim Kolár, Ralf-Peter Tornow, Jirí Jan, Attila Budai, Markus A. Mayer, Martina Vodakova, Robert Laemmer, Martin Lamos, Zdenek Kuna, Jirí Gazárek, Tomas Kubena, Pavel Cernosek, and Marina Ronzhina. Thickness related textural properties of retinal nerve fiber layer in color fundus images. *Computerized Medical Imaging and Graphics*, 38(6):508–516, 2014.

[21] Harry A. Quigley and Aimee T. Broman. The number of people with glaucoma worldwide in 2010 and 2020. *British Journal of Ophthalmology*, 90(3):262–267, 2006.

[22] Chandrasekaran Raja and Narayanan Gangatharan. A hybrid swarm algorithm for optimizing glaucoma diagnosis. *Computers in Biology and Medicine*, 63:196–207, 2015.

[23] Suresh Sundaram, R. Venkatesh Babu, and H. J. Kim. No-reference image quality assessment using modified extreme learning machine classifier. *Applied Soft Computing*, 9(2):541–552, 2009.

[24] Fengshou Yin, Jiang Liu, Damon Wing Kee Wong, Ngan Meng Tan, Carol Yim-lui Cheung, Mani Baskaran, Tin Aung, and Tien Yin Wong. Automated segmentation of optic disc and optic cup in fundus images for glaucoma diagnosis. In *25th IEEE International Symposium on Computer-Based Medical Systems (CBMS)*, pages 1–6, 2012.

[25] Runxuan Zhang, Guang-Bin Huang, N. Sundararajan, and P. Saratchandran. Multicategory classification using an extreme learning machine for

microarray gene expression cancer diagnosis. *IEEE/ACM Transactions on Computational Biology and Bioinformatics*, 4(3):485–495, 2007.

[26] Zhuo Zhang, Jiang Liu, Damon Wing Kee Wong, Ngan Meng Tan, Joo-Hwee Lim, Shijian Lu, Huiqi Li, Ziyang Liang, and Tien Yin Wong. Neuro-retinal optic cup detection in glaucoma diagnosis. In *2nd IEEE International Conference on BioMedical Engineering and Informatics (BMEI)*, pages 1–4, 2009.

8

Early Diagnosis of Diabetes using an Intelligent Machine Learning Technique

C. V. Guru Rao[1] and Nilgün Şengöz[2]

[1]Department of Computer Science and Engineering Gayatri Vidhya Parishad College of Engineering, Visakhapatnam, India
[2]Department of Information Systems and Technologies, Gölhisar School of Applied Sciences, Burdur Mehmet Akif Ersoy University, Turkey
E-mail: guru_cv_rao@hotmail.com; nilgunsengoz@mehmetakif.edu.tr

Abstract

Medical research on diabetes prediction is essential. Several models are available for use in making diagnoses of diabetes in patients. However, due to the low quality, we were unable to provide a useful outcome. More noise features can be found in the genetic data, increasing the already high difficulty of diabetes prediction. The negatives caused poor prediction and efficiency ratings. Thus, the purpose of the proposed work is to create a unique Coati-based multilayer model (CBMM) for determining whether or not a patient would develop diabetes. Preprocessing, feature selection, categorization, and gene expression are just some of the tasks that have been completed. At the outset, we imported the data sensed by IoT devices, preprocessed the data, and chose relevant attributes. The next step is to classify the patients' health statuses based on your predictions. At last, the model's efficacy was evaluated, and it was found to have a very high prediction exactness score. Therefore, the developed method has been tried and evaluated in Python, where it has achieved the maximum score for predicting gene expression and the lowest miss classification rate.

Keywords: Genomic data, gene expression, multilayer model, Coati optimization, noise filtering, feature analysis.

8.1 Introduction

Diabetes mellitus (DM) is a significant public health concern in the twenty-first century. It is defined by elevated blood glucose levels (hyperglycemia) and is associated with long-term complications that can lead to serious damage in many organs such as the heart, kidneys, feet, and peripheral nerves of the retina. In the current period characterized by significant technical advancements, it is noteworthy that a definitive and enduring solution for diabetes diseases remains elusive within the contemporary society [1]. According to a recent study, it has been discovered that approximately 50% of individuals who have recently been diagnosed with diabetes exhibit clinical symptoms of both microvascular and macrovascular diseases. Hence, the timely identification of diabetes is of utmost importance in order to prevent its continuous advancement, as well as the development of related problems. Consequently, there is a need for cost-effective approaches to diagnose diabetes. The successful execution of this task necessitates the utilization of several methodologies and information systems [2]. In order to undertake a therapeutic intervention for the mitigation of this pathological condition within the human organism, it is imperative to implement rigorous surveillance measures. The management of diabetes can be accomplished by the implementation of many strategies, including the administration of medication and insulin, engaging in regular physical activity, adhering to a balanced diet, monitoring blood glucose levels, and seeking frequent medical guidance. This intervention is expected to contribute to the management and preservation of diabetes mellitus [3]. This condition is characterized by its ability to elevate glucose or sugar levels in the circulatory system. Typically, there is a fluctuation in the concentration of glucose under normal daily circumstances. It is imperative to implement vigilant eye monitoring for each patient. The compromised interaction between insulin and glucose results in the accumulation of significant amounts of glucose inside the bloodstream [4]. Inadequate therapy can result in elevated glucose levels, hence exacerbating the poor state. The global mortality rate is significantly impacted by the complexities associated with diabetic diseases, leading to loss of life among individuals [5]. In order to mitigate risk, individuals diagnosed with diabetes can effectively manage their health status. A rigorous surveillance system is implemented in order to mitigate the potential hazards associated with this perilous ailment. The early detection of diabetes is a complex task that requires the utilization of various criteria, including predictive, quantitative, and systematic methodologies [6]. The utilization of insulin and a nutritious

8.1 Introduction

diet might enhance energy levels in combating this ailment. Diabetes has been found to significantly contribute to the development of various health conditions, including but not limited to heart disease, respiratory infections, stroke, and a range of other ailments [7]. The healthcare facilities own a substantial database with a large number of individuals diagnosed with diabetes. There are various classifications of diabetes conditions [8].

Insufficient insulin production within blood cells can have implications for immune system functioning and the impairment of pancreatic function within the body [9]. Individuals diagnosed with diabetes experience metabolic dysfunction as a result of diminished insulin secretion and/or impaired insulin action. Insulin plays a crucial role in the regulation of blood glucose levels. Cardiovascular disease is commonly observed in individuals with diabetes as the prevailing manifestation [10]. Over the past decade, there has been a global increase of 70% in mortality rates among those with diabetes. There is a significant demand for the implementation of machine learning (ML) [32, 33] techniques in the early detection of diabetes [11]. The utilization of artificial intelligence [34, 36] and machine learning (ML) [35] models inside the medical diagnosis system, specifically developed for early detection of diabetic symptoms, along with comprehensive analysis of laboratory tests and disease-related information, has proven to be highly advantageous in mitigating potential harm to patients [12]. Researchers these days are focusing on the different kinds of databases to figure out how to predict diabetes signs. The suggested method is needed to find the early stages of the disease and check the blood sugar level often [13]. This method guarantees a good way of life for surveying in this world. Not having enough of any one of the 40 genes can lead to genetic diabetes [14]. It takes a lot of computing power to use ML-based models to make the early stages of forecast work better [15]. These days, this sickness also affects kids. Genetic diseases, bad eating habits, being overweight, and smoking are all things that people have in common [16].

The proposed machine learning techniques based on intelligence aim to develop a model that can acquire and enhance its knowledge and experience during the initial phase of the diagnostic process [17, 37]. Analyzing the input data provided by the algorithm is useless for understanding the model [18]. Because of this, it is helpful to make accurate predictions and decisions based on the early stages of identification [19]. The initial stage involves data preparation, followed by preprocessing and addressing the relevant disease, and subsequently eliminating extraneous data from the database [20]. Assume that the large amount of learning data makes making a judgment difficult. As a result, the designed algorithm generates logic functions, probability checks, statistics analysis, and so on [21]. The model must undertake data analysis and extract insights based on experiential knowledge. The accuracy of the performance analysis is calculated, and optimization is conducted using the fresh dataset [22]. The algorithm facilitates the processes of disease tracing, segmentation, and classification, followed by an assessment of disease severity. The model that has been developed is founded upon the machine learning methodology for comprehending the patterns and knowledge pertaining to individuals with diabetes, and its efficacy is assessed through performance evaluation.

The key contribution of this study is defined as follows:
- Initially, the genomic data is collected and trained to the Python system as the input data.
- Moreover, a novel Coati-based multilayer model (CBMM) was executed with the optimal feature analysis module.
- Then, the trained data were preprocessed, and the error-free data was given to the input as the following layer.
- Consequently, the genetic changes were predicted and matched with the diabetes genetic features. Based on the matching rate, the diabetes occurrence possibility was predicted.
- Finally, the performance metrics like accuracy, recall, precision, f-score, and error rate were measured, and improvement scores than the existing models were noted.

The present work paper is presented in the form of related work in the second section. The limitations of the standard technique are discussed in Section 8.3. Then the solution to the problem specified is elaborated in Section 8.4. The validation result of the novel solution is discussed in Section 8.5. The research work carried out in this chapter is concluded in Section 8.6.

8.2 Related Works

Few recent related studies are explained as follows:

Jiheum Park et al. [23] created a deep learning framework to predict pancreatic cancer risk using longitudinal EHR data. A new training process, a "grouped" neural network (GrpNN) architecture, leverages representation learning to build a dimensionally reduced vector for each measurement set before making a prediction. They were models. Using EHR data from Columbia University Irving Medical Center-New York Presbyterian Hospital, their framework outperformed logistic regression, XGBoost, and feedforward neural networks in early detection (AUROC 0.671, CI 95% 0.667−0.675, $p < 0.001$) at 12 months before diagnosis. It showed that their masking method improves distal times before diagnosis and that their GrpNN model reduces overfitting relative to the feedforward baseline, improving generalizability. Results were consistent by race. Their technique could be used to other disorders including cancer, where early detection improves survival.

Maria Tariq et al. [24] discuss publicly available diabetic retinopathy datasets, and our model is trained and tested on Kaggle. Here, image preparation and data augmentation are key to handling noisy and small datasets. To address the uneven nature of diabetic retinopathy datasets, recent data augmentation strategies are reviewed. For Kaggle dataset discovery and imbalanced classification, the proposed approach uses deep learning and reinforcement learning.

For COVID-19 metagenomic next-generation sequencing (mNGS) samples, Fatma Hilal Yagin et al. [25] developed an explainable artificial intelligence (XAI) model using machine learning. COVID-19 genes were selected using LASSO. LR, SVM, RF, and XGBoost were used to predict COVID-19. An explainable technique using LIME and Shapley additive explanations (SHAP) approaches was used to identify COVID-19-associated biomarker candidate genes and improve model interpretability. LIME indicated that high IFI27 gene expression increased positive class probability. The XGBoost model predicted COVID-19. Machine learning, LIME, and SHAP can explain COVID-19 biomarker prediction and give doctors an intuitive grasp and interpretability of risk factors in the model.

Amjad Rehman Khan et al. [26] present a deep learning approach to classify brain tumors using an MRI data analysis to assist practitioners. The recommended method comprises three main phases: preprocessing, brain tumor segmentation using k-means clustering, and, finally, classify tumors into their respective categories (benign/malignant) using MRI data through a fine-tuned VGG19 (i.e., 19 layered Visual Geometric Group) model. The proposed approach was evaluated on BraTS 2015 benchmarks datasets through rigorous experiments. The results endorse the effectiveness of the proposed strategy and it achieved better accuracy compared to the previously reported state-of-the-art techniques.

Bala Anand Muthu et al. [27] presented a wearable sensor for IoT-based healthcare data mining research. To acquire patient data from IoT, we design generalized approximate reasoning-based intelligence control (GARIC) with regression rules. Finally, utilize Boltzmann belief network deep learning to train the data for AI. GWAS is then used to predict illnesses. Thus, if people have ailments, they will receive SMS, email, and other warnings before receiving treatment and advice from doctors. The prediction rate is greatest. Real-time implementation is not possible using the approach.

8.3 System Model and Problem Statement

It is fair to say that diabetes is among the incurable diseases. Therefore, the severity of diabetes could be managed by forecasting its traits at an earlier stage. Prediction of diabetes using imaging methods and other numerical models is already feasible. But those algorithms cannot pinpoint a precise severity spectrum. In light of these concerns, the DNA data was analyzed to diagnose the disease in its earliest stages. Additionally, several intelligent neural procedures were established for assessing the genomic data; however, the usual prediction features in the neural system are insufficient for analyzing the illness characteristics. Due to the gravity of the problem, our work

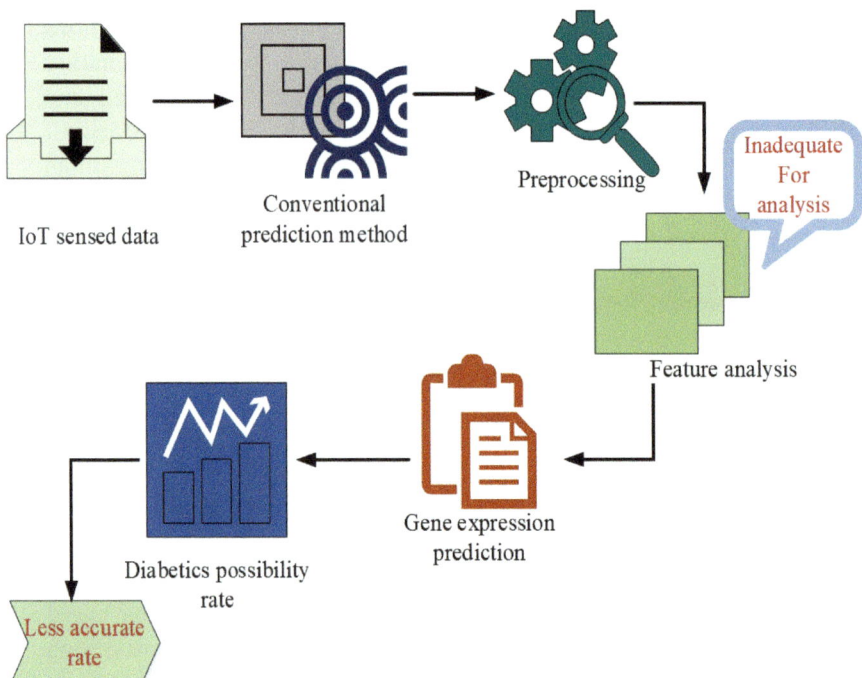

Figure 8.1 Difficulties in standard prediction method.

was inspired to develop a reliable predictive mechanism using cutting-edge techniques like neural networks and optimum algorithms.

Figure 8.1 depicts the categorization scheme used by the conventional prediction method. The standard method's feature analysis over-optimizes the complexity of the algorithm and over-extracts irrelevant characteristics, leading to an erroneous gene prediction rate. Methodology for implementing a Coati-based multilayer model for evaluating gene expression is presented as a means of overcoming these challenges.

8.4 Proposed Model

For this purpose of genomic feature analysis, a new Coati-based multilayer model (CBMM) is presented. Here, the early projection was carried out based on the unique human's genetic alterations. Therefore, the likelihood of developing diabetes could be estimated with more precision based on the

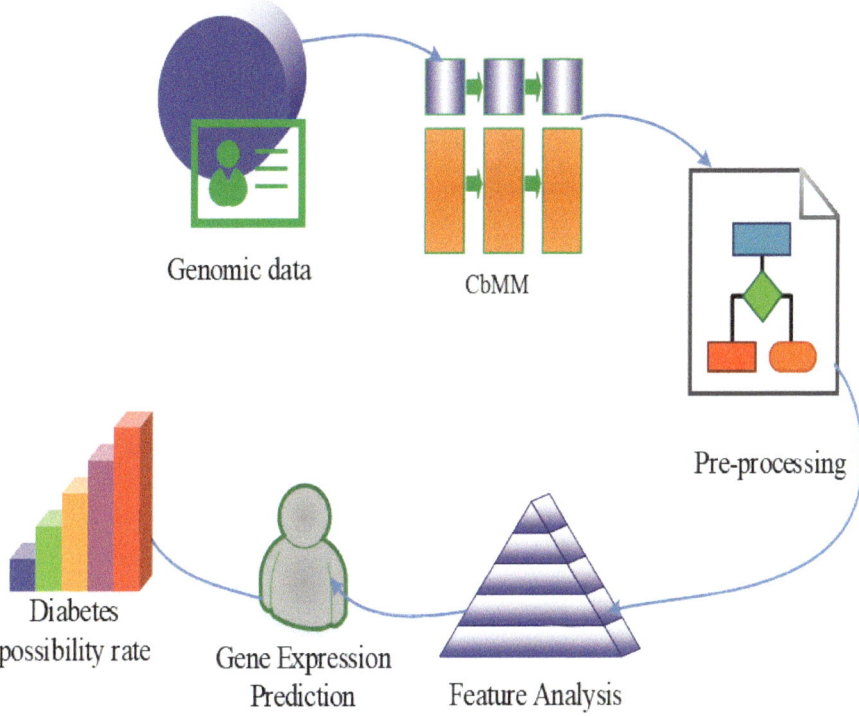

Figure 8.2 Proposed architecture.

variation in genetic traits. Finally, the prediction resilience was evaluated using f-score, Accuracy, precision, error rate, and recall, and compared to other recent related works. Figure 8.2 depicts the suggested process flow.

In Figure 8.2, we see the architecture of the proposed technique laid out. The goal of this research is to use genetic information to detect diabetes at an early stage. By comparing studies, one may see how much progress has been made in predicting genes across all databases.

8.4.1 Process of the proposed methodology

Input, Hidden, Classification, Optimization, and Output Layers make up the five-stage processing architecture of the proposed technique. The multilayer perceptron and the Coati optimization algorithm can also be used to complete the task. The multilayer perceptron is a computational element of a neural network simulation. The best filtering result can be achieved with the help of the deep, smart multilayer perceptron model. The CBMM's hidden layer performs the filtering operation in this case.

Figure 8.3 details the many processing stages involved in the CBMM approach. In this case, the input layer received the collected dataset, while the hidden layer performed the noise filtering. It is possible to collect the data after noise removal and then incorporate it into the classification stage. The output layer then received the optimized result after the classification parameters were changed using the Coati optimization function.

8.4.1.1 Data training and preprocessing

In the beginning stage, the genomic data was gathered and imported. Eqn (8.1) can describe the data training function.

$$F(Gn(d)) = Gn(d)\{1, 2, 3, 4, \cdots, N\}, \tag{8.1}$$

where Gn denotes the gene expression, data was designated as d, and $Gn(d)$ represents the data of gene expression. Noise was preprocessed or filtered out using the machine learning approach, which reduced the algorithm complexity during testing and training. Therefore, preprocessing is clearly a crucial step.

$$P(Gn(d)) = nx - (d)[(nx, lx)]. \tag{8.2}$$

The data is the most important thing that made the model training possible because machine learning methods depend on it. The sample that was collected has both normal and a lot of noise features. So, preprocessing is

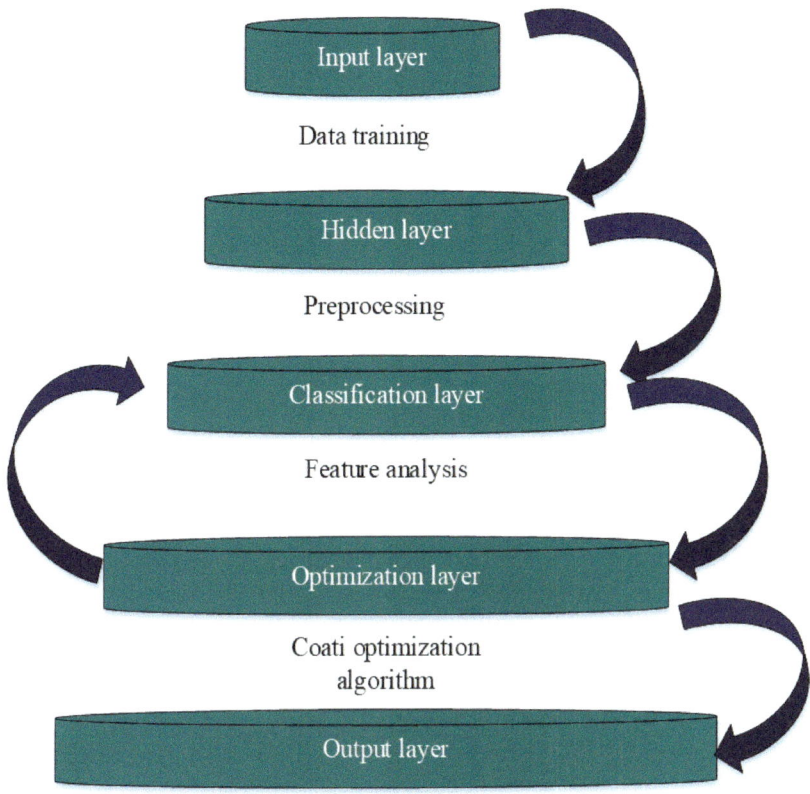

Figure 8.3 Processing layers of CBMM.

needed to get rid of all the noise and make a clean collection. Eqn (8.2) does the preprocessing. Here P is the preprocessing variable. The noise features can be denoted as $nxlx$ and can be represented as normal features. The output of the preprocessing phase is the noise-free dataset.

8.4.1.2 Feature analysis

The procedure of feature analysis has been conducted in order to minimize computational expenses. Feature analysis refers to the systematic examination and extraction of the most relevant characteristics associated with a particular disease from a comprehensive set of trained datasets. The prediction and analysis of gene expression have been conducted on a specific set of test data, resulting in a high accuracy score. The dataset contains both maximum matching features and minimum matching features.

$$F(Gn(d)) = f_{min} + C(0,1) * (f_{max} - f_{min}). \tag{8.3}$$

The feature analyzing parameter is denoted as F. The current maximum matched features can be represented as f_{max} and the minimum matched features can be represented as f_{min}. Hence, the minimum matching features of the disease can be eliminated from the maximum matching features of the disease. Thus, the maximum matching features can be extracted. The class is denoted c, and the disease class features can be represented as $(0,1)$. The feature analysis mathematical expression can be denoted in eqn (8.3).

8.4.1.3 Classification and prediction

The utilization of a classification model is necessary in order to discern between normal data and disease-related aspects. Typically, the initial dataset has a combination of anomalous and normal traits. Therefore, based on the disease data, it is possible to identify the gene expression patterns that are associated with the disease characteristics. Ultimately, the classification layer has been utilized to categorize the illness aspects.

$$S(\text{disease\_feature}) = \begin{cases} \text{if}(Gn=0) & \text{Normal} \\ \text{else} & \text{Abnormal} \end{cases}, \tag{8.4}$$

where the parameters of the disease were set in the form of $(0,1)$. Processing the condition in the testing phase can be in the form of if($Gn[0,1]$). The classification of the disease features is described in eqn (8.4), where 0 indicates the normal features, and 1 indicates the abnormal features.

8.4.1.4 Gene expression analysis

After completing the purpose of the feature analysis and classification phase, predict the tested disease samples in the current gene features. The gene expression prediction can be derived by using eqn (8.5).

$$\text{Probability\_exp} = 1 \times \frac{dt}{\max\_Gene} + 0.1, \tag{8.5}$$

where 1 is the probability score representing the analyzed maximum possible current genes. In addition, 1 is the probability representing 100% in the tested samples of the present gene. The tested disease samples are denoted as dt and the max_Gene represents some crucial features like blood cholesterol levels, glucose, proteins, etc. Based on the genomic database, the features of the gene are different.

Algorithm: 1 CBMM

Start
{
    Dataset initialization()
    {
        $int\ \mathrm{Gn}(d) = 1, 2, 3, \cdots n;$
        //initialize the dataset
    }
    Preprocessing()
    {
        $int\ P, nx, lx;$
        //initialize the preprocessing variables
        $P(Gn(d)) \longrightarrow \mathrm{Remove} - \mathrm{nx}(\mathrm{Gn}(d))$
        //Eliminating the noise features
    }
    Feature analysis()
    {
        $int\ f_{\min}, f_{\max}, F;$
        //initialize the feature extracting variables
        $\mathrm{Extract} \longrightarrow f_{\max}(\mathrm{Gn}(d))$
        //extracting the maximum matched features
    }
    Classification and prediction()
    {
        if($\mathrm{Gn}(d) = 0$)
        {
        Normal
        }else **Abnormal**
    Gene expression()
    {
        $\mathrm{Gene} - \mathrm{probability} \longrightarrow \mathrm{Classification}(\mathrm{max\_gene})$
        //classified the maximum number of genes
    }
}
Stop

The gradual process of the advanced methodology is described in Figure 8.4. The designed pseudo code of the detailed mathematical formulation was represented in Algorithm 1. After the preprocessing of the

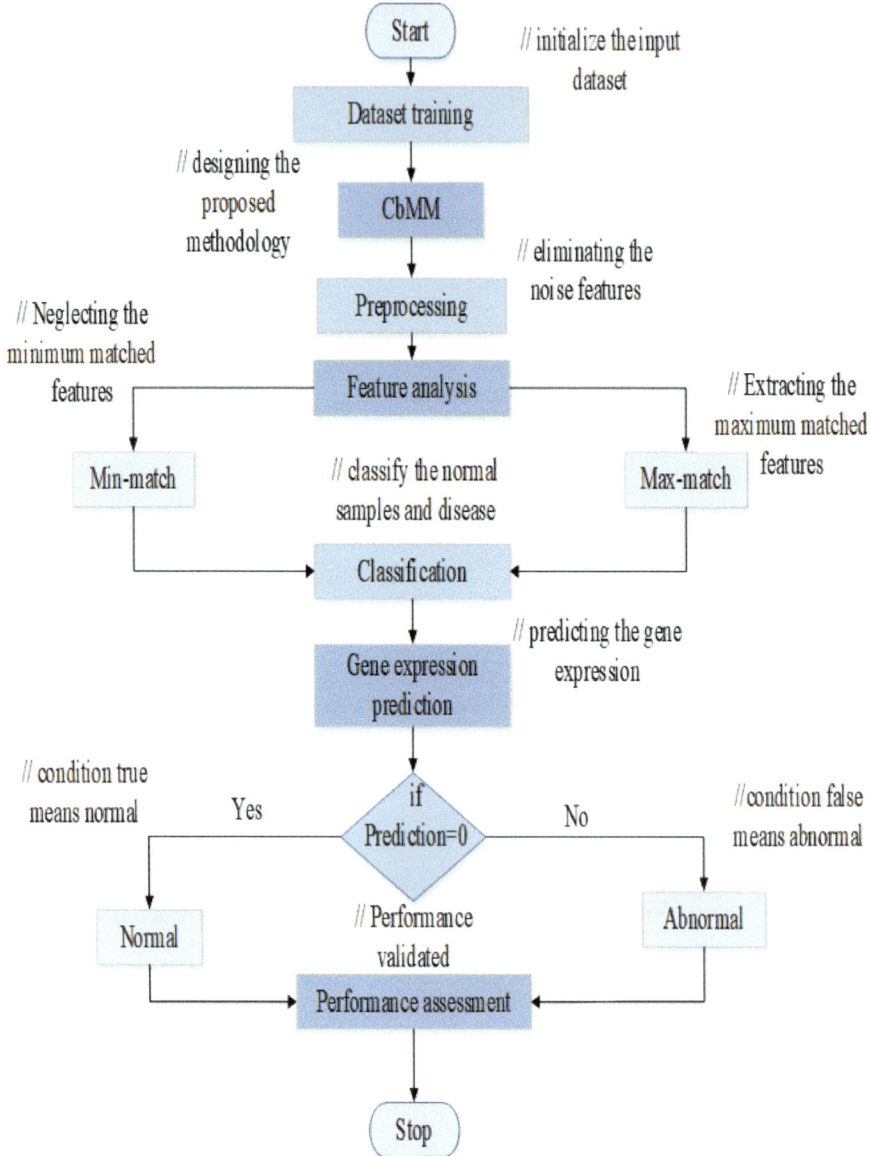

Figure 8.4 The flow diagram of CBMM.

designated steps, the different classification measures have esteemed the performance of the novel CBMM.

8.5 Result and Discussion

The novel CBMM methodology is executed in the Python programming platform and runs on the Windows 10 platform. In the beginning, the data was collected and trained in the system. The data contains both abnormal and normal data features.

The specification of the implementation parameters is described in Table 8.1. The preprocessing phase removed the noise features and imported the error-free features to the classification phase. In addition, the classification phase predicted the gene expression and measured the performance.

8.5.1 Case study

To check the performance of the proposed methodology, some of the test validation was executed, and the results are represented systematically. For the test validation, the genomic dataset was adopted. It contains medical and demographic data. The dataset comprises a total of 15,485 samples. In the total samples, 12,388 are the training samples, and 3097 are the testing samples.

Table 8.2 describes the details of the database. Moreover, samples are considered for training, in that 6209 are the normal samples and 6179 are the abnormal samples. Also, the samples considered for testing are 3097, in that 1519 are the normal samples and 1578 are the abnormal samples.

Figure 8.5 represents the accuracy and loss assessment of the CBMM over the training epoch. The assessment of the diabetes prediction system's reliability is conducted by evaluating its train and test validation accuracy score. The failure ratio of the implemented framework is determined using loss metrics, which are recorded throughout both the training and test validation phases.

The outcome of the predicted result was exposed as a confusion matrix in Figure 8.6. The categorization result was obtained as positive and negative

Table 8.1 Execution parameters specification.

Description of parameters	
Programming environment	Python
Database	Genomic data
Dataset format	Numerical data
Operating system	Windows 10
Deep Network	Multilayer network
Optimization	Coati optimization

Table 8.2 Database details.

Total number of samples: 15,485	
Normal	7728
Abnormal	7757
Training (80%): 12,388	
Normal	6209
Abnormal	6179
Testing (20%): 3097	
Normal	1519
Abnormal	1578

scores for both true and false categories. In this scenario, the forecast was categorized into two distinct scenarios, denoted as 0 and 1. The value of 0 represents the standard or expected condition, while a value of 1 signifies an atypical or deviant state.

8.5.2 Performance assessment

The performance of the suggested methodology was evaluated by computing metrics such as f-score, precision, accuracy, and recall to validate its effectiveness. To assess the enhancement in performance, consider the model that has been recently affiliated. The existing models are Long-Short Term Memory

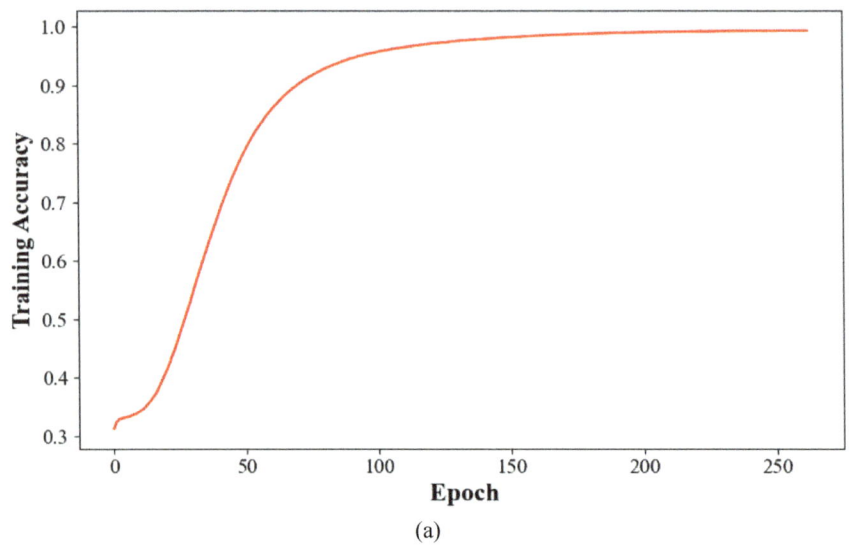

(a)

Figure 8.5 *Continued.*

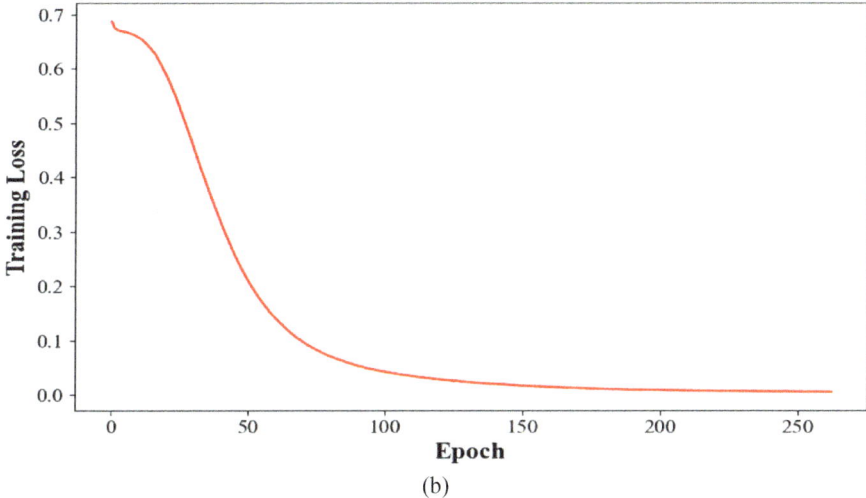

Figure 8.5 (a) Training accuracy. (b) Training loss.

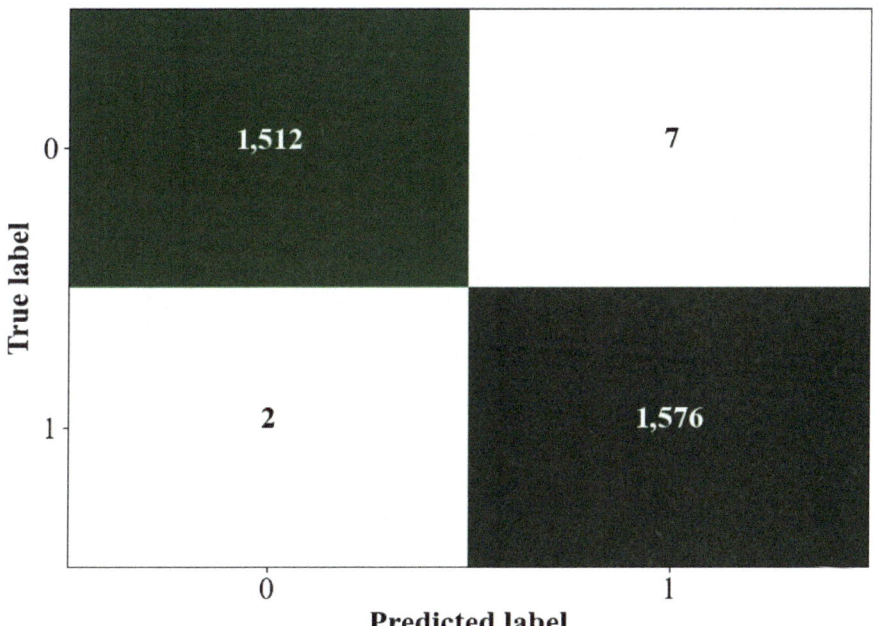

Figure 8.6 Confusion matrix.

(LSTM) [28], random forest (RF) [29], soft voting classifier (SVC) [30], and K-nearest neighbor (KNN) [31].

8.5.2.1 Precision
Precision refers to the quantification of the correct positive predictions made throughout the disease classification procedure. The precision parameter is developed in eqn (8.6).

$$\Pr ecision = \frac{T_p}{F_p + T_p}. \qquad (8.6)$$

The precision score of the LSTM method is 98.3%, the SVC method earned 73.13%, RF gained 80%, and the KNN gained 88.29%. Considering the compared mechanism, the proposed novel CBMM methodology achieved 99.70% precision. The statistics are revealed in Figure 8.7.

8.5.2.2 Accuracy
The accuracy of the metrics measures how well the gene expression predicts the range. To figure out how accurate a prediction is, you take the average of the numbers that were negative and positive. The accuracy of the gene

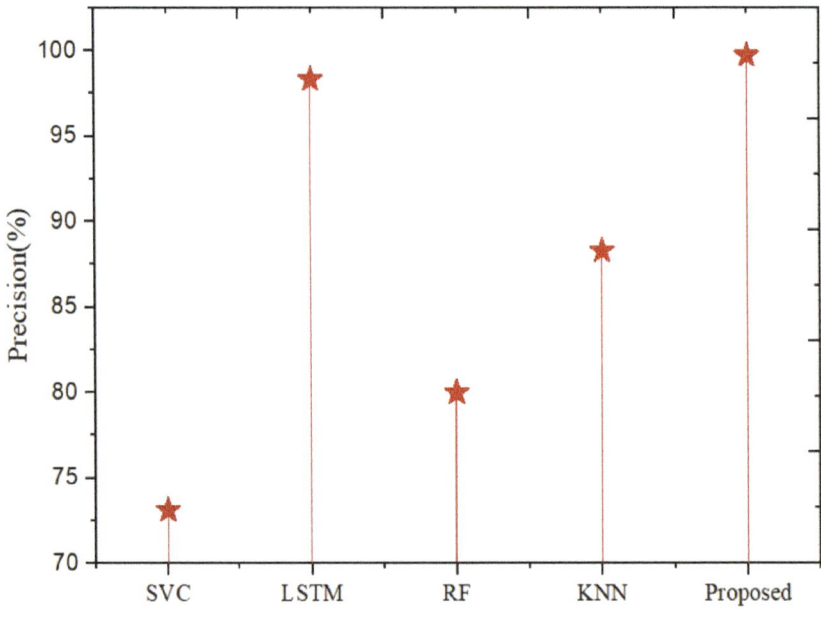

Figure 8.7 Precision assessment.

prediction can be equated by using eqn (8.7).

$$\text{Accuracy} = \frac{T_n + T_p}{T_p + F_p + T_n + F_n}. \tag{8.7}$$

The accuracy score of the LSTM method is 92.3%, the SVC method earned 79.08%, RF gained 86.36%, and the KNN gained 87.61%. Considering the compared mechanism, the proposed novel CBMM methodology earned 99.70% accuracy. The statistics are revealed in Figure 8.8.

8.5.2.3 Recall
The metrics recall is confirmed with false and true scores to calculate the stability range of successful prediction. The recall parameter is measured by using eqn (8.8):

$$\text{Recall} = \frac{T_p}{T_p + F_n}. \tag{8.8}$$

The recall score of the LSTM method is 83.8%, the SVC method earned 70%, RF gained 98%, and the KNN gained 73.45%. Considering

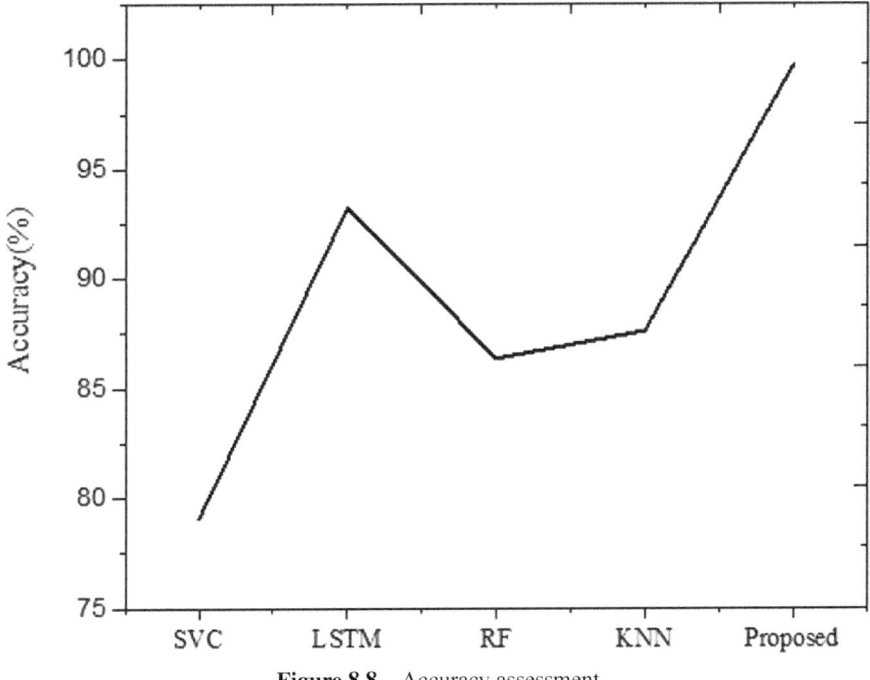

Figure 8.8 Accuracy assessment.

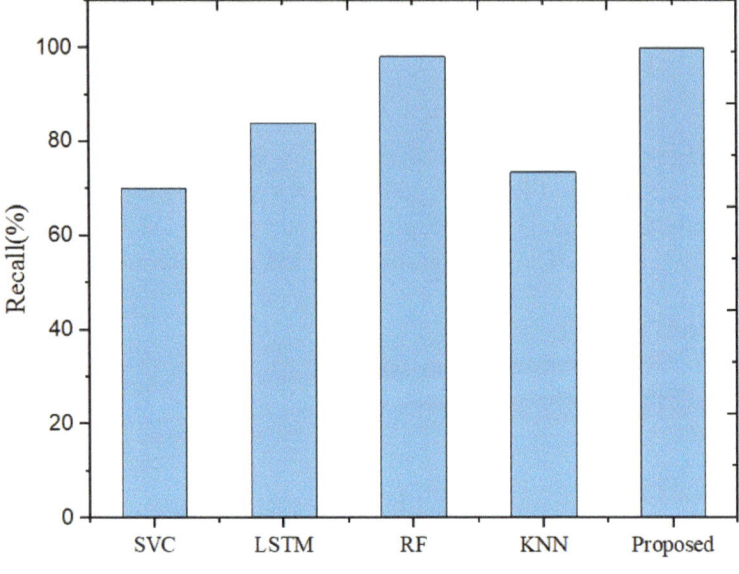

Figure 8.9 Recall assessment.

the compared mechanism, the proposed novel CBMM methodology earned 99.70% recall. The statistics are revealed in Figure 8.9.

8.5.2.4 F-score

The recall metrics value the mean range of the false and true scores. Therefore, the f-score is the average prediction of the true and false scores. Eqn (8.9) can describe the f-score.

$$\text{f} - \text{score} = 2 \times \frac{\text{recall} \times \text{precision}}{\text{precision} + \text{recall}} \quad (8.9)$$

The f-score earned by the LSTM method is 90.5%, the SVC method gained 71.56%, RF earned 88%, and the KNN earned 80.19%. Considering the compared mechanism, the proposed novel CBMM methodology earned a 99.70% f-score. The statistics are revealed in Figure 8.10.

8.5.2.5 Error rate

Error rate metrics measure the prediction error degree of a methodology with respect to the true methodology. The error rate can be measured by using eqn (8.10)

$$\text{Error rate} = \frac{F_p + F_n}{T_p + T_n + F_p + F_n}. \quad (8.10)$$

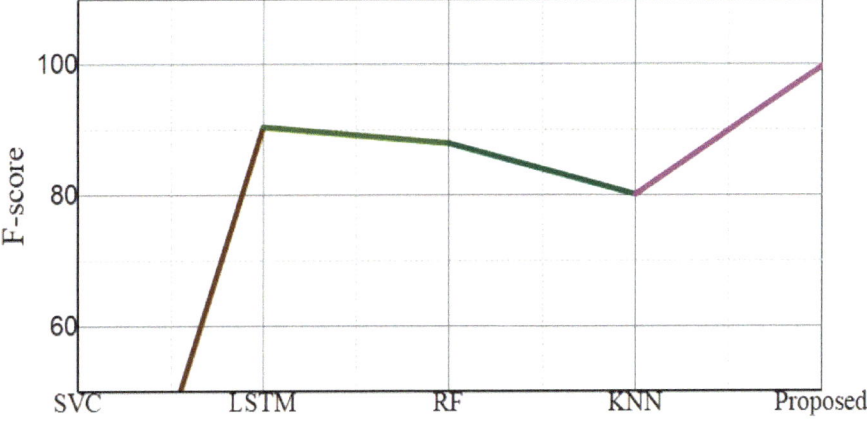

Figure 8.10 F-score assessment.

The error rate of the LSTM method is 0.08, the SVC method is 0.21, the RF method is 0.14, and the KNN method is 0.13. Considering the compared mechanism of the proposed novel CBMM methodology, the error rate is 0.0029. The statistics are revealed in Figure 8.11.

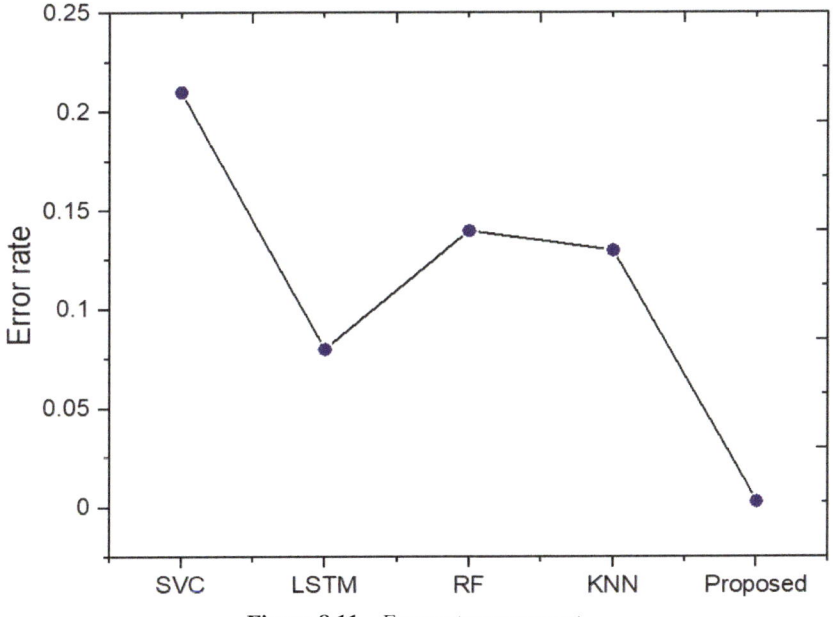

Figure 8.11 Error rate assessment.

Table 8.3 Comparison assessments.

Methods	Accuracy (%)	Precision (%)	Recall (%)	F-score (%)	Error rate
LSTM	92.3	98.3	83.8	90.5	0.08
SVC	79.08	73.13	70	71.56	0.21
RF	86.36	80	98	88	0.14
KNN	87.61	88.29	73.45	80.19	0.13
Proposed	99.70	99.70	99.70	99.70	0.0029

Table 8.4 Overall performance of CBMM.

Performance of CBMM	
Efficiency parameters	Performance (%)
Accuracy	99.70
Precision	99.70
Recall	99.70
f-score	99.70
Error rate	0.0029
Time	21.9302 seconds

8.5.2.6 Time

Time required to complete the overall process of diabetes prediction is time taken for training plus time taken for testing. In the proposed methodology, the time taken for the training is 15.69 seconds, the testing time is 0.0119 seconds, and the execution time is 21.9302 seconds.

8.5.3 Discussion

Overall, the presented model has obtained a better metrics score in all metrics: accuracy, precision, recall and f-score. Table 8.3 presents such results. In addition, the overall performance of the designed novel CBMM has been tabulated in Table 8.4.

The overall performance of the novel CBMM methodology is described in Table 8.4. All the forecasting metrics have reported 99.70% recognition effectiveness. The procedure under consideration achieved an accuracy rate of 99.70%. The accuracy rate demonstrates a notable improvement in comparison to the currently employed strategy. The suggested model achieved the optimal outcome in the diabetes prediction parameters, demonstrating the great performance of the unique CBMM.

8.6 Conclusion

In the current work, the researchers employed the CBMM methodology, a novel approach, to predict diabetes in individuals using genomic data. The genomic dataset was utilized for the purpose of test validation. The dataset comprises medical and demographic information. At the onset, the noise characteristics are removed from the dataset. Next, the desired features should be chosen in the feature selection function, while the undesired traits are disregarded. Subsequently, the input for the classification layer consisted of the data that had been verified to be free of errors. The Coati optimization approach is employed for feature selection, while the multilayer perceptron is utilized for predicting diabetes, with the accuracy score being the evaluation metric. The forecast encompasses two scenarios, namely the standard case and the atypical case. Subsequently, the performance of the methodology was evaluated by utilizing the categorization results. The accuracy of the diabetes prediction algorithm is 99.70%. The prediction accuracy has been enhanced by 1% in comparison to the conventional model. Additionally, the proposed work exhibits an error rate of 0.0029%. In comparison to the conventional approach, there was a 2% enhancement in the score. The methodology that was suggested demonstrated a significant level of performance in predicting diabetes. Nevertheless, the current work lacks the incorporation of security measures. In the future, incorporating the security implementation in conjunction with the suggested model is expected to yield improved outcomes along with the support of artificial intelligence [37].

Acknowledgments

Writing a book chapter is harder than I thought and more rewarding than I could have ever imagined. This work is not possible without the cooperation of my colleague Dr. Geetha Manoharan, Assistant Professor, School of Business, SR University and her research scholar Ms. Sunitha at School of Business.

I am also thankful to my all colleagues for their editing help, late-night feedback sessions, and moral support. I would like to extend my sincere thanks to my research assistants, librarian, and study participants who impacted and inspired me.

Finally, I would be remiss in not mentioning my family, especially my spouse and children. Their belief in me has kept my spirit and motivation high during this process.

References

[1] Abdelsalam MM. Effective blood vessel reconstruction methodology for early detection and classification of diabetic retinopathy using OCTA images by an artificial neural network. Informatics in Medicine Unlocked. 2020 Jan 1;20:100390.

[2] Alrefai N, Ibrahim O, Shehzad HM, Altigani A, Abu-ulbeh W, Alzaqebah M, Alsmadi MK. An integrated framework-based deep learning for cancer classification using microarray datasets. Journal of Ambient Intelligence and Humanized Computing. 2023 Mar;14(3): 2249-60.

[3] Vij R, Arora S. A novel deep transfer learning based computerized diagnostic Systems for Multi-class imbalanced diabetic retinopathy severity classification. Multimedia Tools and Applications. 2023 Mar 3:1-38.

[4] Sarki R, Ahmed K, Wang H, Zhang Y, Ma J, Wang K. Image preprocessing in classifying and identifying diabetic eye diseases. Data Science and Engineering. 2021 Dec;6(4):455-71.

[5] Mohanty A, Parida S, Nayak SC, Pati B, Panigrahi CR. Study and impact analysis of machine learning approaches for smart healthcare in predicting mellitus diabetes on clinical data. Smart Healthcare Analytics: State of the Art. 2022:75-101.

[6] Quazi S. Artificial intelligence and machine learning in precision and genomic medicine. Medical Oncology. 2022 Jun 15;39(8):120.

[7] Tamal M, Alshammari M, Alabdullah M, Hourani R, Alola HA, Hegazi TM. An integrated framework with machine learning and radiomics for accurate and rapid early diagnosis of COVID-19 from Chest X-ray. Expert systems with applications. 2021 Oct 15;180:115152.

[8] Peiffer-Smadja N, Dellière S, Rodriguez C, Birgand G, Lescure FX, Fourati S, Ruppé E. Machine learning in the clinical microbiology laboratory: has the time come for routine practice? Clinical Microbiology and Infection. 2020 Oct 1;26(10):1300-9.

[9] Muhammad LJ, Algehyne EA, Usman SS, Ahmad A, Chakraborty C, Mohammed IA. Supervised machine learning models for prediction of COVID-19 infection using epidemiology dataset. SN computer science. 2021 Feb;2:1-3.

[10] Awotunde JB, Ayo FE, Jimoh RG, Ogundokun RO, Matiluko OE, Oladipo ID, Abdulraheem M. Prediction and classification of diabetes mellitus using genomic data. InIntelligent IoT systems in personalized health care 2021 Jan 1 (pp. 235-292). Academic Press.

[11] Goecks J, Jalili V, Heiser LM, Gray JW. How machine learning will transform biomedicine. Cell. 2020 Apr 2;181(1):92-101.
[12] Acs B, Rantalainen M, Hartman J. Artificial intelligence as the next step towards precision pathology. Journal of internal medicine. 2020 Jul;288(1):62-81.
[13] Papadimitroulas P, Brocki L, Chung NC, Marchadour W, Vermet F, Gaubert L, Eleftheriadis V, Plachouris D, Visvikis D, Kagadis GC, Hatt M. Artificial intelligence: Deep learning in oncological radiomics and challenges of interpretability and data harmonization. Physica Medica. 2021 Mar 1;83:108-21.
[14] Lu J, Song E, Ghoneim A, Alrashoud M. Machine learning for assisting cervical cancer diagnosis: An ensemble approach. Future Generation Computer Systems. 2020 May 1;106:199-205.
[15] Ghaffar Nia N, Kaplanoglu E, Nasab A. Evaluation of artificial intelligence techniques in disease diagnosis and prediction. Discover Artificial Intelligence. 2023 Jan 30;3(1):5.
[16] Battineni G. Machine Learning and Deep Learning Algorithms in the Diagnosis of Chronic Diseases. Machine Learning Approaches for Urban Computing. 2021:141-64.
[17] Joshi I, Bhrdwaj A, Khandelwal R, Pande A, Agarwal A, Srija CD, Suresh RA, Mohan M, Hazarika L, Thakur G, Hussain T. Artificial intelligence, big data and machine learning approaches in genome-wide SNP-based prediction for precision medicine and drug discovery. InBig Data Analytics in Chemoinformatics and Bioinformatics 2023 Jan 1 (pp. 333-357). Elsevier.
[18] Khan S, Nazir S, García-Magariño I, Hussain A. Deep learning-based urban big data fusion in smart cities: Towards traffic monitoring and flow-preserving fusion. Computers & Electrical Engineering. 2021 Jan 1;89:106906.
[19] Oh SH, Lee SJ, Park J. Effective data-driven precision medicine by cluster-applied deep reinforcement learning. Knowledge-Based Systems. 2022 Nov 28;256:109877.
[20] Adlung L, Cohen Y, Mor U, Elinav E. Machine learning in clinical decision making. Med. 2021 Jun 11;2(6):642-65.
[21] Kabir MF, Chen T, Ludwig SA. Performance analysis of dimensionality reduction algorithms in machine learning models for cancer prediction. Healthcare Analytics. 2023 Nov 1;3:100125.
[22] Comito C, Falcone D, Forestiero A. Convergence Between IoT and AI for Smart Health and Predictive Medicine. InIntegrating Artificial

Intelligence and IoT for Advanced Health Informatics: AI in the Healthcare Sector 2022 Feb 24 (pp. 69-84). Cham: Springer International Publishing.

[23] Park J, Artin MG, Lee KE, Pumpalova YS, Ingram MA, May BL, Park M, Hur C, Tatonetti NP. Deep learning on time series laboratory test results from electronic health records for early detection of pancreatic cancer. Journal of Biomedical Informatics. 2022 Jul 1;131:104095.

[24] Tariq M, Palade V, Ma Y, Altahhan A. Diabetic Retinopathy Detection Using Transfer and Reinforcement Learning with Effective Image Pre-processing and Data Augmentation Techniques. InFusion of Machine Learning Paradigms: Theory and Applications 2023 Feb 7 (pp. 33-61). Cham: Springer International Publishing.

[25] Yagin FH, Cicek İB, Alkhateeb A, Yagin B, Colak C, Azzeh M, Akbulut S. Explainable artificial intelligence model for identifying COVID-19 gene biomarkers. Computers in Biology and Medicine. 2023 Mar 1;154:106619.

[26] Khan AR, Khan S, Harouni M, Abbasi R, Iqbal S, Mehmood Z. Brain tumour segmentation using K-means clustering and deep learning with synthetic data augmentation for classification. Microscopy Research and Technique. 2021 Jul;84(7):1389-99.

[27] Muthu B, Sivaparthipan CB, Manogaran G, Sundarasekar R, Kadry S, Shanthini A, Dasel A. IOT based wearable sensor for diseases prediction and symptom analysis in the healthcare sector. Peer-to-peer networking and applications. 2020 Nov;13:2123-34.

[28] Refat MA, Al Amin M, Kaushal C, Yeasmin MN, Islam MK. A comparative analysis of early-stage diabetes prediction using machine learning and deep learning approach. In2021 6th International Conference on Signal Processing, Computing and Control (ISPCC) 2021 Oct 7 (pp. 654-659). IEEE.

[29] Hassan MM, Peya ZJ, Mollick S, Billah MA, Shakil MM, Dulla AU. Diabetes prediction in healthcare at an early stage using machine learning approach. In2021 12th International Conference on Computing Communication and Networking Technologies (ICCCNT) 2021 Jul 6 (pp. 01-05). IEEE.

[30] Kumari S, Kumar D, Mittal M. An ensemble approach for classification and prediction of diabetes mellitus using the soft voting classifier. International Journal of Cognitive Computing in Engineering. 2021 Jun 1;2:40-6.

[31] Reddy DJ, Mounika B, Sindhu S, Reddy TP, Reddy NS, Sri GJ, Swaraja K, Meenakshi K, Kora P. Predictive machine learning model for early detection and analysis of diabetes. Materials Today: Proceedings. 2020 Oct 23.

[32] Geetha Manoharan & Dr. Barinderjit Singh Dr. Y. Saritha Kumari, Dr. Aarti Dawra, Dr. Rachana Jaiswal, Aijaj Ahmad Raj, An evaluation of machine learning techniques and how they affect human resource management and sustainable development, Manager - The British Journal of Administrative Management, ISSN: 1746 – 1278, pp 4-14, September2022. https://tbjam.org/vol58-special-issue-06

[33] Barinderjit Singh & Dr. Rachana Jaiswal Habibulla Palagiri, Dr. Shweta Mogre, Dr. Rashmi Badjatya Rawat, Geetha Manoharan, an investigation on the use of machine learning methods for predicting employee performance, Manager - The British Journal of Administrative Management, ISSN: 1746 – 1278, pp 15-22, September2022. https://tbjam.org/vol58-special-issue-06/

[34] Geetha Manoharan, Dr. Harish Purohit Dr. Somanchi Hari Krishna, Dr. Neelam Sheoliha, Shivangi Ghildiyal, Dr. R. Krishna Vardhan Reddy, An overview of exploring the potential of artificial intelligence approaches in digital marketing, Manager - The British Journal of Administrative Management, ISSN: 1746 – 1278, pp 48-59, September2022. https://tbjam.org/vol58-special-issue-06/

[35] Tripathi, M. A., Tripathi, R., Effendy, F., Manoharan, G., Paul, M. J., & Aarif, M. (2023, January). An In-Depth Analysis of the Role That ML and Big Data Play in Driving Digital Marketing's Paradigm Shift. In 2023 International Conference on Computer Communication and Informatics (ICCCI) (pp. 1-6). IEEE.

[36] Manoharan, G., Durai, S., Ashtikar, S. P., & Kumari, N. (2024). Artificial Intelligence in Marketing Applications. In Artificial Intelligence for Business (pp. 40-70). Productivity Press.

[37] Abdulwahid, A. H., Pattnaik, M., Palav, M. R., Babu, S. T., Manoharan, G., & Selvi, G. P. (2023, April). Library Management System Using Artificial Intelligence. In 2023 Eighth International Conference on Science Technology Engineering and Mathematics (ICONSTEM) (pp. 1-7). IEEE.

[38] Hegde N, Krishna S, Manvi SS. Diabetic Retinopathy Diagnosis System Based on Artificial Intelligence. In Human-Machine Interface Technology Advancements and Applications 2023 Sep 22 (pp. 213-230). CRC Press.

[39] R. P. Kumar, S. S. Vandana, D. Tejaswi, K. Charan, R. Janapati and U. Desai, "Classification of SSVEP Signals using Neural Networks for BCI Applications," 2022 International Conference on Intelligent Controller and Computing for Smart Power (ICICCSP), Hyderabad, India, 2022, pp. 1-6.

[40] Vijayvargiya A, Singh B, Kumar R, Desai U, Hemanth J. Hybrid Deep Learning Approaches for sEMG Signal-Based Lower Limb Activity Recognition. Mathematical Problems in Engineering. 2022 Nov 26;2022.

9

Advanced Diabetes Prediction: A Comprehensive Analysis of Machine Learning and Deep Learning Techniques

Thottempudi Pardhu[1], Anwar Bhasha Pattan[1], Vijay Kumar[2], Usha Desai[3], and Biswaranjan Acharya[4]

[1]Department of ECE, BVRIT Hyderabad College of Engineering for Women, India
[2]SENSE, Vellore Institute of Technology, India
[3]Department of Electronics and Communication Engineering, S.E.A College of Engineering & Technology, Bengaluru, India
[4]Department of Computer Engineering - AI & BDA, Marwadi University, India
E-mail: pardhu.t@bvrithyderabad.edu.in; vijaykumar@vit.ac.in; dr.ushadesai@seaedu.ac.in; biswaacharya@ieee.org

Abstract

Diabetes is a common disease in the world. Recently, various machine learning (ML) methods employed to predict diabetes incidence have been presented. This topic's vast complexity motivates research on various deep learning (DL) approaches. So far, the best accuracy achieved among all such methods is 95.1%, which was achieved using the model combined with CNN-LSTM. Several machine learning techniques have attempted to solve this problem, but a group of classifiers still needs to be employed. This makes it attractive to assess the effectiveness of the prediction of diabetes. Furthermore, no updated survey must be assessed to compare the effectiveness of all proposed machine learning (ML) and deep learning (DL) techniques and their ensemble models. The present study conducted a thorough, all-inclusive analysis of the diabetes prediction studies presented over the last six years based on the ML and DL approaches. In another research, the Pima Indian

Dataset experimented with other rarely used classifiers of machine learning to see whether, due to implementation, any noticeable performance came out or not. The result presented a range of 68%–74% due to the implementation of those classifiers. These classifiers should be applied in diabetes prediction and improved by creating integrated models.

Keywords: Diabetes prediction, machine learning, deep learning, CNN-LSTM.

9.1 Introduction

Diabetes has emerged as a significant health concern among older populations globally. According to the International Diabetes Federation's 2017 statistics, approximately 451 million people were diagnosed with diabetes, a number expected to increase to 693 million by 2045. Defined as a chronic condition, diabetes manifests through abnormal glucose levels in the blood, often due to pancreatic issues that result in inadequate insulin production or its complete absence, characteristic of type-1 diabetes. In other cases, it stems from the body's resistance to insulin, commonly leading to type-2 diabetes. The root causes of diabetes remain under debate, but it is widely accepted that genetic predispositions and environmental lifestyle factors are influential. While diabetes is typically a lifelong condition, it can be controlled through effective management and medication. It also carries the risk of secondary health complications, such as cardiovascular diseases and neuropathy. Early detection and proactive diabetes management are crucial in minimizing these risks and averting more severe health issues. Notably, researchers in bioinformatics are investigating predictive models for diabetes employing various machine learning algorithms, including decision trees, support vector machines (SVM), and linear regression.

Artificial neural networks (ANNs), known for their robust performance, have long been a staple in machine learning. However, the emergence of deep learning (DL), a more advanced iteration of ANNs, has been a game-changer, especially with the exponential growth in data volume and complexity. Recent experiments using DL techniques have shown promising results, sparking further research to enhance these models' accuracy through innovative or integrated approaches. The Pima Indian Dataset from the UCI repository is a popular choice for research on diabetes prediction, serving as a benchmark for the performance of these advanced models.

While traditional machine learning and deep learning methods have been extensively used in diabetes prediction, a dearth of studies focusing on cutting-edge deep learning techniques is evident. Many studies have explored the use of decision trees, support vector machines, ANNs, and specific DL strategies. Some have concentrated on machine learning classifiers for predicting diabetes, while others have examined machine learning applications across various diseases, including diabetes. However, the number of studies specifically investigating deep learning techniques for diabetes prediction is limited, and some have confined their scope to this disease alone, neglecting a broader range of conditions.

This chapter delves into machine learning and deep learning methodologies, including combination models, as applied in diabetes prediction research until 2013. Combination models, or ensemble classifiers, merge multiple machine learning techniques. The analysis reviews various classifiers introduced over the past six years, concentrating on machine learning and profound learning developments. Due to article length restrictions, only a select group of studies is discussed here, with a detailed list in Tables 9.1, 9.2, and 9.3.

Table 9.1 Overview of machine learning algorithm applications.

Author	Year	Technique	Result	Dataset
Kandhasamy et al. [4]	2015	J48	Accuracy: 73.85%	Pima Indians Diabetes Database
		K-nearest neighbors (KNN)	Accuracy: $K = 1$, 70.2%; $K = 3$, 72.7%; $K = 5$, 73.2%	
		Random forest	Accuracy: 71.8%	
Iyer et al. [5]	2015	J48	Accuracy: 74.91%	Pima Indians Diabetes Database
		Naïve Bayes	Accuracy: 77.01%	
Yuvaraj et al. [17]	2017	Random forest	Accuracy: 94.05%	Pima Indians Diabetes Database
		Decision tree (ID3)	Accuracy: 88.25%	
		Naïve Bayes	Accuracy: 91.50%	
Tafa et al. [18]	2015	SVM	Accuracy: 95.77%	Private dataset (collected manually)
		Naïve Bayes	Accuracy: 94.85%	
Sisodia et al. [19]	2018	Decision tree	Accuracy: 73.95%	Pima Indians Diabetes Database
		SVM	Accuracy: 65.60%	
		Naïve Bayes	Accuracy: 77.30%	
Mercaldo et al. [20]	2017	J48	Precision: 0.84, recall: 0.84, F-measure: 0.84	Pima Indians Diabetes Database
		MLP neural network	Precision: 0.79, recall: 0.79, F-measure: 0.79	

Table 9.1 *Continued.*

Author	Year	Technique	Result	Dataset
		HoeffdingTree	Precision: 0.79, recall: 0.79, F-measure: 0.79	
Negi et al. [21]	2016	SVM	Accuracy: 73.93%	Global dataset combined of all the available datasets
Olaniyi et al. [22]	2014	Multilayer feed-forward neural network back-propagation algorithm (ANN)	Accuracy: 84%	Pima Indians Diabetes Database
Soltani et al. [23]	2016	Probabilistic neural network (PNN)	Training Accuracy: 90.56%, Testing Accuracy: 83.49%	Pima Indians Diabetes Database
Somnath et al. [24]	2017	Two-class neural network	Accuracy: 85.3%	Pima Indians Diabetes Database
Mamuda et al. [25]	2017	Levenberg–Marquardt learning algorithm	MSE: 0.00025060	Pima Indians Diabetes Database
		Bayesian regulation learning algorithm	MSE: 2.023e-05	
		Scaled conjugate gradient learning algorithm	MSE: 6.3583	
Kumari et al. [26]	2013	SVM	Accuracy: 80%	Pima Indians Diabetes Database
Farran et al. [27]	2013	Logistic regression (LR), K-NN, SVM	Accuracy: LR 80.4%, K-NN 79.6%, SVM 80.4%	Kuwait Health Network (KHN)
Tapak et al. [28]	2013	ANN, SVM, RF	AUC: ANN 76.1%, RBF-SVM 98.9%, RF 77.3%	Iran Population Dataset
Anand et al. [29]	2013	Higher order neural network (HONN)	MSE: training 6.5257e-04, testing 0.4219e-05	Pima Indians Diabetes Database

Table 9.1 *Continued.*

Author	Year	Technique	Result	Dataset
Choi et al. [30]	2014	ANN, SVM	Dataset for KNHANES 2010: Accuracy: ANN 65%, SVM 66.9%	KNHANES 2010 and 2011 Datasets
			Dataset for KNHANES 2011: Accuracy: ANN 63.7%, SVM 67.1%	
Sarwar et al. [31]	2014	Naïve Bayes	Accuracy: 98%	Private dataset (collected manually)
		ANN	Accuracy: 97%	
		K-NN	Accuracy: 93%	
Durairaj et al. [32]	2015	Back-propagation neural network and Levenberg–Marquardt optimizer	Accuracy: 93%	Pima Indians Diabetes Database
Anand et al. [33]	2015	CART	Accuracy: 77%	Private dataset (collected manually)
Malik et al. [34]	2016	LR, SVM, ANN	Accuracy: LR 77.86%, RBF-SVM 86.09%, NN 82.7%	Private dataset (collected manually)
Perveen et al. [35]	2016	Standalone J48, Adaboost ensemble using J48, bagging ensemble using J48	AROC: bagging ensemble using J48 0.97%	CPCSSN Database
Joshi et al. [36]	2016	Back-propagation neural network (ANN)	Accuracy: 83%	Not mentioned
Sowjanya et al. [37]	2015	J48	Sensitivity: 0.92, Specificity: 0.92	Private dataset (collected manually)
		Naïve Bayes	Sensitivity: 0.78, Specificity: 0.89	
		SVM with polykernel	Sensitivity: 0.85, Specificity: 0.91	
		SVM with RBF kernel	Sensitivity: 0.88, Specificity: 0.88	
		MLP neural network	Sensitivity: 0.85, Specificity: 0.91	

Table 9.1 Continued.

Author	Year	Technique	Result	Dataset
Cai et al. [38]	2015	LR, LDA, Naïve Bayes, SVM	mRMR method: SVM 77%, LR: 77%, LDA: 77%, NB: 73%	Chinese Gut Microbiota Datasets
			mRMR method: SVM: 72%, LR: 74%, LDA: 69%, NB: 70%	European Gut Microbiota Datasets
Maniruzzaman et al. [39]	2017	Kernel cross-validation 10: linear discriminant analysis	Accuracy: 79.86%	Pima Indians Diabetes Database
		Kernel cross-validation 10: quadratic discriminant analysis	Accuracy: 78.56%	
		Kernel cross-validation 10: Naïve Bayes	Accuracy: 79.57%	
		Kernel cross-validation 10: Gaussian process	Accuracy: 83.97%	
Mirshahvalad et al. [40]	2017	Perception	Accuracy: 0.75	NHANES0506, NHANES0708, and NHANES0910
		Ensemble perception	Accuracy: 0.79	
Sun et al. [41]	2017	Kernel-based adaptive filtering algorithm	CGM signals prediction accuracy	Private dataset

The study also identifies the most frequently used ML classifiers and explores less common ones using the Weka tool and the Pima Indian Dataset (PID). This approach aims to fill gaps in current research regarding these classifiers. The findings are compared with other studies using the same dataset, with the ultimate goal of offering a centralized database for diabetes prediction research.

The document is structured as follows. Section 9.2 overviews previous studies utilizing ML and DL techniques, including hybrid models. Section 9.3 presents a detailed analysis of relevant studies, describing primary diabetes datasets and summarizing their characteristics (Table 9.4). It also briefly overviews various ML/DL algorithms, highlighting their strengths and

Table 9.2 Comprehensive review of deep learning technique implementations.

Authors	Year	Technique	Result	Dataset
Mohebbi et al. [43]	2016	Unsupervised deep learning neural network (deep patient)	AUC–ROC: 0.93	Electronic health records
Sun et al. [42]	2017	Convolutional neural network (CNN)	Accuracy: 79.5%	Continuous glucose monitoring (CGM) signals
		Multilayer perceptron neural network (MLP)	Accuracy: 74.5%	
		Logistic regression	Accuracy: 67.2%	
Ashiquzzaman et al. [7]	2017	Recurrent deep neural network (RNN)	Accuracy: type-1 diabetes = 79%, type-2 diabetes = 83%	Pima Indians Diabetes Dataset
Miotto et al. [44]	2017	Deep neural network long short-term memory (LSTM)	Precision: 60.6%	Large regional Australian hospital dataset
		Markov chain neural network	Precision: 35.1%	
		Plain recurrent deep neural network (RNN)	Precision: 59.0%	
Lekha et al. [46]	2018	Modified convolution neural network (CNN)	ROC: 0.98	Breath Dataset Private
Swapna et al. [8]	2018	Convolutional neural network (CNN)	Accuracy: 92.9%	Electrocardiograms (ECG) private
		Convolutional neural network (CNN) combined with long short-term memory (LSTM)	Accuracy: 96.1%	
Ashiquzzaman et al. [7]	2018	Deep learning architecture (MLP/GRNN/RBF)	Accuracy: 89.41%	Pima Indians Diabetes Dataset

weaknesses (Table 9.5). Section 9.4 showcases a case study on diabetes prediction results and discusses the performance of different classifiers. Finally, Section 9.5 concludes the investigation with key findings and implications.

Table 9.3 Overview of combined model approaches in research.

Authors	Year	Technique	Result	Dataset
Askarzadeh et al. [47]	2013	Feed-forward neural network and bird mating optimizer	Training: average 25.57, std. dev. 1.60; testing: average 23.57, std. dev. 2.89	Pima Indians Diabetes Database
Nirmala Devi et al. [11]	2013	Simple KNN	Accuracy: 75.17%	Pima Indians Diabetes Database
		K-means and KNN	Accuracy: 97.5%	
		Amalgam and KNN	Accuracy: 97.8%	
Rahimloo et al. [9]	2016	Logistic regression and feed-forward neural network	Error rate: 0.0007	Association of Diabetic's City of Urmia Dataset
Gill et al. [10]	2016	Support vector machine (SVM) and neural network (NN)	Accuracy: 96.59%	Pima Indians Diabetes Database
Soltani et al. [23]	2017	Classification and regression trees (CART) for Fuzzy Rule Generation in diabetes prediction	Accuracy: 93%	Pima Indian Dataset, Mesothelioma, WDBC, StatLog, Cleveland, Parkinson's Telemonitoring
Rao et al. [48]	2018	Sequential minimal optimization (SMO), SVM and elephant herding optimizer	Accuracy: 79.21%	17 medical datasets including Pima Indian Diabetes Database

9.2 Related Work

This survey presents a thorough compilation of 27 works investigating machine learning. The details of these studies can be found in Table 9.1. The references for these studies range from [4, 5, 17–41]. Given space constraints, only the ten most recent research works will be thoroughly examined. In addition, this section will examine seven research publications on deep learning approaches, as specified in Table 9.2. Table 9.3 mentions six additional papers and integrated models, but this work does not provide further discussion on these.

Tables 9.1 and 9.2 provide crucial information about each study published throughout the past six years. This information includes the reference, publication year, assessment measures, and their corresponding values, as well as the datasets utilized. Additionally, Table 9.4 provides a comprehensive analysis of the datasets used in these investigations, including their sizes and key attributes.

9.2.1 Literature review on diabetes prediction using machine learning

The medical field increasingly recognizes the effectiveness of machine learning (ML) algorithms in predicting diseases, especially diabetes. Kandhasamy and Balamurali used a dataset from the UCI repository to test various classifiers, such as SVM, J48, K-nearest neighbors (KNN), and random forest. They evaluated the specificity, sensitivity, and accuracy of these models using a fivefold cross-validation method, both with and without data preprocessing, though details of the preprocessing were not disclosed. The J48 decision tree classifier achieved an accuracy of 73.82% without preprocessing [4]. After applying preprocessing techniques, the random forest and KNN ($k = 1$) reached a perfect accuracy of 100%.

Yuvaraj and Sripreethaa's study focused on the accuracy of the random forest, Naïve Bayes, and decision tree algorithms on a preprocessed dataset of Pima Indian Diabetes. The dataset was split into 30% for testing and 70% for training, with the random forest model delivering an impressive 94% accuracy rate [17]. Tafa and colleagues introduced a novel approach by developing a combined model of support vector machines (SVM) and Naïve Bayes to predict diabetes. Their study utilized a unique dataset from Kosovo, which included 402 individuals, 80 of whom had type-2 diabetes. The dataset featured diverse characteristics, including dietary habits, physical activity levels, and family history. The combined model demonstrated a remarkable accuracy of 97.6%, surpassing the individual accuracies of SVM and Naive Bayes [18].

Deepti and Dilip examined the Pima Indian Dataset using a decision tree, SVM, and Naive Bayes classifier with a tenfold cross-validation approach. Specific data preprocessing techniques were not mentioned. They measured F-measure, accuracy, precision, and recall, with Naive Bayes achieving the % accuracy rate of 76.30% [19]. Mercaldo et al. analyzed the Pima Indian Dataset using six classifiers (J48, Multilayer Perceptron, HoeffdingTree, JRip, BayesNet, and RandomForest). Critical traits were identified using the

GreedyStepwise and BestFirst algorithms without any preprocessing. The HoeffdingTree technique achieved the highest metrics, with a precision of 0.757, a recall of 0.762, and an F-measure of 0.759 [20].

Negi and Jaiswal researched diabetes prediction using SVM on a combined dataset of Pima Indians and Diabetes 130-US. The dataset, containing 102,538 samples with 49 different features, underwent preprocessing, replacing missing or out-of-range values, converting non-numeric variables into numeric forms, and standardizing the data. Various feature selection strategies yielded a 72% accuracy rate [21]. Olaniyi and Adnan's research showcased a high level of accuracy, predicting diabetes with an 82% success rate. They achieved this using a multilayer feed-forward neural network with back-propagation on the normalized Pima Indian Diabetes database, which was divided into 500 training and 268 testing samples [22].

Soltani and Jafarian applied a probabilistic neural network (PNN) to the unprocessed Pima Indian Dataset, dividing it into 90% training and 10% testing. Their accuracies were 89.56% and 81.49%, respectively [23]. Rakshit and colleagues used a two-class neural network on the preprocessed and standardized Pima Indian Dataset. Feature selection was based on correlation, with the dataset divided into 314 training and 78 testing samples, achieving an accuracy of 83.3% [24]. Mamuda and Sathasivam utilized tenfold cross-validation to apply three supervised learning algorithms — Levenberg–Marquardt (LM), Bayesian regulation (BR), and scaled conjugate gradient (SCG) — on the Pima Indian Dataset. The LM method demonstrated exceptional performance, indicated by a mean squared error (MSE) of 0.00024091 [25].

9.2.2 Literature review on diabetes prediction using deep learning

Researchers have recognized the efficacy of deep learning (DL) methods for handling large datasets, particularly for predicting diabetes. Seven notable studies have been conducted in the past six years, detailed in Table 9.2. Ashiquzzaman et al. [7] introduced an innovative approach using a deep neural network (DNN) that integrates a multilayer perceptron (MLP), a general regression neural network (GRNN), and a radial basis function (RBF). They chose not to preprocess the data to leverage the DNN's natural data filtering capabilities. The model was trained and tested using the Pima Indian database, divided into 192 test samples, achieving an accuracy of 88.41%.

Swapna and colleagues [8] explored deep learning's high accuracy potential. They evaluated CNN and CNN-LSTM models on an electrocardiogram dataset comprising 142,000 samples across eight variables. Employing five-fold cross-validation without preprocessing, these models achieved accuracies of 90.9% and 95.1%, respectively, as indicated in Table 9.4.

Mohebbi et al. [42] evaluated logistic regression, a multilayer perceptron neural network, and a conventional neural network (CNN) using continuous glucose monitoring (CGM) data from nine patients, simulating 97,200 CGM days. Utilizing a leave-one-patient-out cross-validation method, the CNN model achieved a notable accuracy of 77.5%, highlighting the complexity of the task given the limited data. Miotto et al. [43] developed "Deep Patient," an unsupervised deep neural network analyzing 704,857 patient records. They recommended using preprocessing techniques like principal component analysis (PCA) to enhance prediction accuracy, achieving an AUC of 0.91. However, the effectiveness of this model depends on the assumption that data is representative and unbiased, which may only sometimes be the case in real-world healthcare settings.

Pham et al. [44] applied three deep learning techniques – long short-term memory (LSTM), Markov, and plain RNN – to data from an Australian hospital, reducing the sample size from 12,000 to 7191 through preprocessing. The LSTM model achieved the highest precision rate of 59.6%. Ramesh et al. [45] used a recurrent neural network (RNN) to predict both types of diabetes using the Pima Indian Dataset, consisting of 768 samples with eight prioritized variables. The dataset was split 80% for training and 20% for testing, with accuracy rates of 78% for type-1 and 81% for type-2 diabetes using holdout validation.

Lekha and Suchetha [46] employed a one-dimensional adapted convolutional neural network (CNN) to predict diabetes from breath signals, using data from 11 healthy individuals, nine with type-2 diabetes, and five with type-1 diabetes, applying leave-one-out cross-validation and achieving an ROC curve performance of 0.96. This study underscores the potential of using non-invasive methods like breath signals for diabetes prediction, demonstrating deep learning's broad applicability in healthcare. Additionally, researchers have developed hybrid models combining machine learning classifiers with artificial intelligence optimizers to enhance prediction accuracy, as detailed in Table 9.3 [9–11, 23, 47, 48]. These models have shown exceptional accuracy in diabetes prediction, suggesting that integrating various methodologies could significantly improve early diagnosis and treatment strategies for diabetes.

9.3 Discussion

9.3.1 Datasets

Diabetes is a widespread condition, yet there are only a few publicly accessible datasets for its prediction. Given the profound influence of datasets on research outcomes, it is essential to conduct a detailed analysis of the primary datasets highlighted in this study, focusing specifically on their features. Table 9.4 briefly summarizes these primary datasets, including their sizes and characteristics.

The primary datasets discussed in related literature include the electrocardiograms (ECG) [8], a proprietary dataset from three sites in Kosovo [18], the Breath Dataset [46], a dataset from the UCI repository [4], and the widely used Pima dataset, featured in over 10 studies reviewed in this survey. These datasets vary in sample size but typically share similar features such as BMI, diastolic blood pressure, plasma glucose concentration, triceps skinfold thickness, diabetes pedigree function, age, family history of diabetes, regular diet, and physical activities. Various feature selection techniques were applied to manage large feature sets. In studies [17, 20, 21], the Pima dataset initially included 13, 8, and 49 attributes, respectively, which were later reduced to 8, 4, and 9 features using feature selection methods. Notably, a study [20] reported reduced precision (0.757), likely due to the fewer features used in the analysis, underscoring the critical role of these characteristics in diabetes prediction. These features were chosen for their proven significance in predicting diabetes, as evidenced in multiple studies [4, 8, 18, 20, 21, 46]. This information is invaluable for researchers aiming to develop new diabetes-related datasets.

9.3.2 Diabetes prediction using machine learning/deep learning techniques

After reviewing six years of research on diabetes prediction, as summarized in Table 9.1, an analysis was conducted to determine how frequently different machine learning classifiers are used. The findings, illustrated in Figure 9.1, show that artificial neural networks (ANN), support vector machines (SVM), decision trees, and Naive Bayes are the most commonly used classifiers for this purpose. Table 9.5 summarizes the pros and cons of these algorithms for a detailed understanding.

Artificial neural networks (ANN) are a versatile tool, drawing inspiration from the human nervous system [49]. They enable learning from direct

Table 9.4 Overview of datasets utilized in diabetes prediction research.

Dataset	Authors	Number of samples	Number of features	Description of features	Link
University of California Irvine (UCI) MACHINE LEARNING REPOSITORY	Kandhasamy et al. [4]	Not specified	8	Features include number of pregnancies, plasma glucose concentration, diastolic blood pressure, triceps skinfold thickness, 2-hour serum insulin, BMI, diabetes pedigree function, age	UCI Diabetes Dataset http://mldata.org/repository/data/view slug/datasets-uci-diabetes/
Pima Indians Diabetes Dataset	Yuvaraj et al. [17]	75,664	13	Common features: Plasma glucose concentration, diastolic blood pressure, triceps skinfold thickness, 2-hour serum insulin, BMI, diabetes pedigree function, age	Kaggle Dataset https://www.kaggle.com/datasets/uciml/pima-indians-diabetes-database
	Sisodia et al. [19]	768	8	Features same as mentioned above	
	Mercaldo et al. [20]	768	8	Features same as mentioned above	
	Negi et al. [21]	102,538	49	Features same as mentioned above	
	Olaniyi et al. [22]	768	8	Features same as mentioned above	
	Soltani et al. [23]	768	8	Features same as mentioned above	
	Somnath et al. [24]	392	8	Features same as mentioned above	
	Mamuda et al. [25]	768	8	Features same as mentioned above	
	Ashiquzzaman et al. [7]	768	8	Features same as mentioned above	
	Balaji et al. [45]	768	8	Features same as mentioned above	

Table 9.4 Continued.

Dataset	Authors	Number of samples	Number of features	Description of features	Link
Electrocardiograms (ECG) private dataset	Swapna et al. [8]	142,000	8	Primary attributes include glucose concentration in plasma, blood pressure, BMI, age	Private
Private dataset collected from three different locations in Kosovo	Tafa et al. [18]	402	8	Includes BMI, pre-meal glucose, post-meal glucose, diastolic blood pressure, systolic blood pressure, family history of diabetes, regular diet, physical activity levels	Not available
Breath dataset	Lekha et al. [46]	15	7	Categories include type (healthy, type 2, type 1), gender, age, BMI, range of HbA1c, duration of diabetes, range of acetone values	Not available

experiences, identification of key features in datasets that may contain irrelevant details, and effective management of ambiguous scenarios [50, 51]. An ANN is composed of three key layers: the input, output, and hidden layers. The hidden layer, made up of neurons, processes data to enhance learning capabilities. The performance of an ANN can be impacted by the number of hidden layers it contains; too many layers might lead to overfitting issues [52]. Various types of ANNs exist, including multilayer perceptrons (MLP), Bayesian neural networks (BNN), and probabilistic neural networks (PNN), each discussed in Table 9.5 regarding their strengths and weaknesses. ANNs have found successful applications in diverse areas such as prediction and forecasting, classification, data correlation and association, robotics, and data filtering [53–55]. The accuracy of diabetes prediction using ANNs has ranged between 60% and 95%.

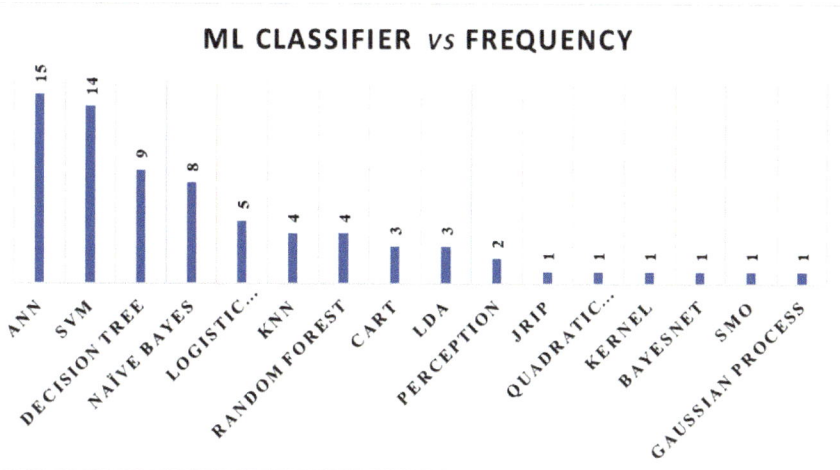

Figure 9.1 Usage frequencies of machine learning classifiers.

A support vector machine (SVM) is a statistical machine learning classifier designed for binary classification tasks. Its unique feature is a hyperplane that separates the two classes [19]. SVMs employ mathematical functions known as kernels to transform incoming data efficiently. These kernels include polynomial, linear, and nonlinear types. Depending on the kernel function used, SVMs are adept at processing unstructured and semi-structured data, such as images and text [18]. While SVMs offer significant advantages, they face challenges such as the complexity of selecting the proper kernel function and the lengthy training process for large datasets. The interpretation of the model's variable weights can also be complex. Despite these issues, SVMs often deliver precise and robust results [34]. They have been extensively used in various medical fields, including diabetes prediction, achieving accuracies between 65% and 96%.

Naive Bayes is a statistical classifier that uses Bayes' theorem to predict the likelihood of an event based on prior knowledge of attribute relationships [19]. It assumes that all features are independent and perform well with datasets that contain missing values or are imbalanced, and it handles both categorical and continuous data effectively. Due to its simplicity and minimal data requirements for setting parameters, it is instrumental in medical diagnostics. However, its assumption of feature independence may affect model accuracy. Naive Bayes has achieved 76%–96% accuracy in diabetes prediction.

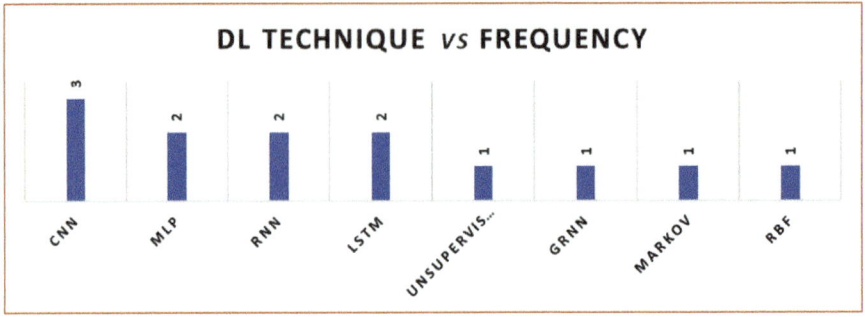

Figure 9.2 Prevalence of deep learning techniques usage.

A decision tree is a supervised learning technique for prediction, classification, and feature selection. The model predicts outcomes based on criteria derived from the dataset [19]. Its advantages include ease of understanding and implementation, especially with the help of visual representations. It effectively handles both numerical and categorical data with minimal preprocessing required. However, decision trees are sensitive to minor data variations, which can significantly affect prediction accuracy [36]. They also become more complex with larger datasets, requiring more time for preparation and analysis [17]. Despite these challenges, decision trees remain a popular choice for diabetes prediction, with accuracy rates ranging from 73% to 88%.

Additionally, deep learning techniques have also been employed in diabetes prediction. The past six years have seen varied applications of methods like convolutional neural networks (CNNs), multilayer perceptron (MLP), recurrent neural networks (RNN), and long-short-term memory (LSTM), as detailed in Table 9.2 and illustrated in Figure 9.2. These approaches, each with their respective strengths and weaknesses, are summarized in Table 9.5.

CNN, a derivative of MLP initially crafted for image processing, comprises three distinct layers: convolutional, pooling, and fully connected layers. It typically utilizes the ReLU activation function [8]. CNN's primary strength lies in its ability to autonomously extract features and identify patterns within data, which is especially beneficial for grid-like data structures [46]. While CNN excels in image processing, it demands significant computational resources and extensive training data to perform optimally [56]. Nonetheless, it offers rapid classification and high accuracy. CNN has been applied in diabetes prediction, achieving 78% and 96% accuracy.

Table 9.5 Comparative analysis of advantages and disadvantages of ML and DL algorithms.

Technique	Advantages	Disadvantages
ANN (artificial neural networks)	(1) Excels in filtering out irrelevant data attributes. (2) Navigates ambiguous scenarios effectively. (3) Demonstrates extensive success across diverse domains, including disease prediction.	(1) Prone to overfitting when using excessive hidden layers. (2) Optimal solutions might not be achieved due to random initialization of weights. (3) Determining structure largely depends on experimentation and experience. (4) Primarily restricted to handling numerical data.
SVM (support vector machines)	(1) Highly efficient with unstructured and semi-structured data such as images and texts. (2) Delivers robust and precise results. (3) Has a strong track record in medical diagnostics and related applications.	(1) Prolonged training durations for extensive datasets. (2) Challenges in choosing the suitable kernel function. (3) Interpreting the significance of variable weights in the model can be complex.
Naive Bayes	(1) Effectively manages datasets with missing elements or imbalances. (2) Accommodates both categorical and numerical data seamlessly. (3) Requires a minimal dataset for configuring parameters.	(1) Operates under the assumption of independent features, which is not always valid. (2) The assumption of conditional independence might compromise accuracy.
Decision tree	(1) Straightforward to comprehend and implement, particularly when visualized graphically. (2) Compatible with both numerical and categorical datasets. (3) Demands minimal preprocessing of data.	(1) High sensitivity to minor data alterations, potentially impacting predictions. (2) Handling large datasets can be challenging due to increased tree complexity.
CNN (convolutional neural networks)	(1) Streamlines the feature extraction process. (2) Effectively discerns correlations within input data. (3) Achieves high accuracy in classification tasks, notably in image recognition.	(1) Involves significant computational resources. (2) Necessitates a large volume of training data for optimal functioning.
MLP (multilayer perceptron)	(1) Adaptable for various tasks including classification, regression, and prediction. (2) Applicable to a wide range of data types, encompassing images and textual information.	(1) Less effective for sophisticated modern computer vision challenges. (2) The potential for a high number of parameters, which can increase processing time.

Table 9.5 *Continued.*

Technique	Advantages	Disadvantages
RNN (recurrent neural networks)	(1) Superior predictive accuracy due to its memory functionality. (2) Aptly addresses complex problem-solving. (3) Capable of processing sequential data, not limited to fixed-size vectors.	(1) Susceptible to the vanishing gradient issue. (2) Encounters challenges in the computation of weights and biases, impacting accuracy.
LSTM (long short-term memory)	(1) Features multiple memory units to tackle the vanishing gradient problem. (2) Predominantly employed in natural language processing (NLP) applications.	(1) High computational demands for memory processing, which can complicate the training process.

MLP, a supervised classifier, features multiple layers of interconnected neurons and aims to create a nonlinear mapping between input and output vectors. Training involves adjusting weights to match the desired output, with the discrepancy between desired and actual outputs serving as the error signal. MLP is suitable for tasks like regression and prediction across various data types, including text and images. Despite its versatility, including predicting ozone levels [57], the fully connected nature of its neurons can lead to a high number of parameters, potentially increasing processing time and limiting its effectiveness in complex computer vision tasks. A detailed analysis of MLP's pros and cons is available in Table 9.5. In diabetes prediction, MLP has reached accuracies from 72.5% to 88.41%, the latter when combined with general regression neural network and radial basis function.

RNN stands out by processing sequential data inputs, such as those in natural language processing, unlike fixed-size vector handling, like in images. A recurrent neural network (RNN) processes input sequences sequentially, retaining information from previous neurons in hidden layers. This allows outputs from various time intervals to inform future neuron inputs. Known for its superior memory capability, RNN excels in complex tasks, including powering Apple's Siri, forecasting text, and language translation [56]. While RNN is proficient in prediction tasks [45], it faces the vanishing gradient problem, where multiplying small numbers can impede weight and bias calculations, diminishing accuracy. RNN's use in diabetes prediction has recorded an accuracy of 81%.

LSTM, an advanced form of RNN, includes memory blocks designed to overcome the issue of vanishing gradients. These blocks feature complex units with one or more memory cells, handling long-term dependencies effectively. Mainly used in natural language processing, LSTM has improved speech recognition, language modeling, and aspects of computer vision. Its main drawback is the extensive computational time required to process memory unit calculations, which complicates training [57]. More details are provided in Table 9.5. In diabetes prediction, LSTM has been used in two ways: combined with CNN, it achieved an accuracy of 95.1%, and when used alone, it reached 59.6%.

9.4 Case Study

9.4.1 Collecting data

The Pima Indian Dataset (PID) was used in this study and is sourced from the UCI Machine Learning Repository. This dataset originated from the National Institute of Diabetes, Digestive, and Kidney Disease [58]. It comprises eight attributes and one output class, represented by a binary value indicating the presence or absence of diabetes. The dataset includes 768 instances: 500 labeled non-diabetic and 268 as diabetic. The PID was chosen for this research due to its broad recognition and its common use as a benchmark dataset for comparing the performance of various methods across different studies.

9.4.2 Preprocessing

Data preprocessing involves preparing and cleaning data before it is used for analysis or modeling. The Weka software was employed in this project to carry out three data preprocessing steps. The attributes in the PID dataset have varying scales, so the values were normalized to a range between 0 and 1. Following normalization, the data was standardized, resulting in a mean of 0 and a standard deviation of 1. This standardization is crucial for Naive Bayes and logistic regression, which assume a Gaussian data distribution. Missing values were then replaced with the mean of the respective attribute. After these preprocessing steps, the data is ready for training using various methodologies. The dataset is divided into training and testing sets through 10-fold cross-validation, ensuring thorough evaluation and validation of the models.

9.4.3 Execution and outcomes

This study analyzed less popular machine learning classifiers and those that have only been used once to predict diabetes in the past six years. The Weka

214 *Advanced Diabetes Prediction: A Comprehensive Analysis of Machine Learning*

software was used to evaluate these classifiers on the Pima Indian Dataset, and their suitability for the dataset guided the selection process. The classifiers were categorized based on their underlying principles, as shown in Figure 9.3. The categories include Trees, Lazy, Rules, Functions, and Bayes. For each classifier, metrics such as accuracy, precision, recall, F-measure, and ROC area were recorded. If these metrics were unavailable, the root mean squared error (RMSE) was used instead, as detailed in Table 9.6.

9.4.4 Analysis and examination

The REPTree classifier achieved the highest accuracy among the classifiers examined: 74.48%. "REP" stands for reduced error pruning, a fast decision tree learning algorithm that generates multiple trees. This classifier is known for its quick learning capability, which results in a short runtime. It operates on the principle of calculating information gain based on entropy.

This concept, exemplified by the M5P model, bridges the gap between actual and desired outputs with remarkable precision levels [59]. The M5P, a high-performance model within the tree category, has been utilized to predict student performance with an astounding precision of 97.17%. In our study,

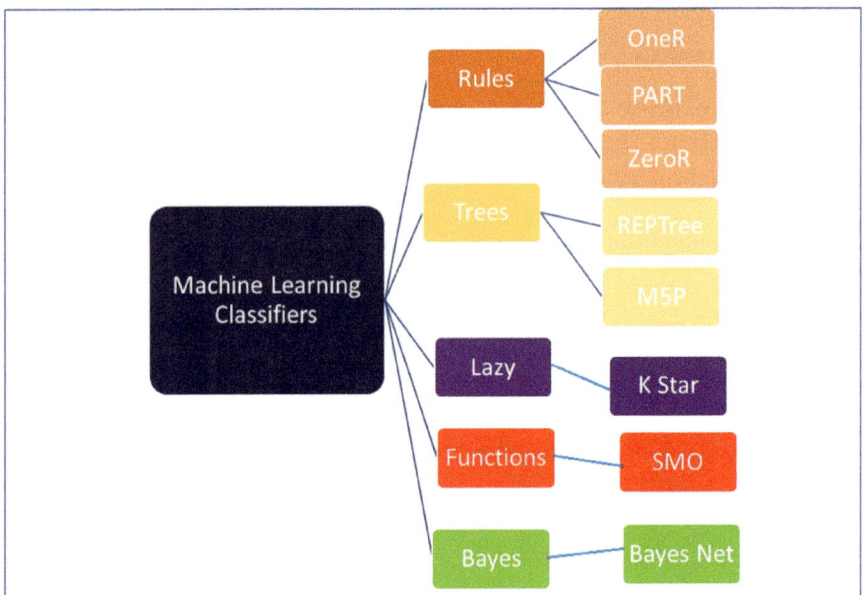

Figure 9.3 Infrequent and underutilized machine learning classifiers in diabetes prediction.

Table 9.6 Performance metrics of infrequently used machine learning classifiers in diabetes prediction.

Classifiers	Accuracy	Precision	Recall	F-Measure	ROC area	Root mean squared error
REPTree	74.51%	0.69	0.54	0.60	0.79	–
M5P	–	–	–	–	–	0.45
KStar	68.25%	0.60	0.36	0.45	0.70	–
oneR	70.95%	0.71	0.48	0.55	0.71	–
PART	74.45%	0.72	0.49	0.60	0.79	–
ZeroR	–	–	–	–	–	0.49
SMO	72.34%	0.80	0.30	0.45	0.65	–
BayesNet	73.93%	0.66	0.60	0.65	0.85	–

the application of this method yielded an impressive root mean squared error of 0.43, a testament to its accuracy. Errors at the nodes of the tree are meticulously quantified, leading to a significant reduction in error rates [60].

Moreover, this technique does not require parameter tuning or prior domain knowledge. The classification process is quick and straightforward, contributing to high accuracy. A recognized limitation of this method is its less optimal performance with non-numeric data. It is also noted that classification error rates tend to increase when working with smaller training datasets [61].

Rule-based classifiers, such as oneR, PART, and ZeroR, have demonstrated accuracies comparable to that of REPTree, instilling confidence in their effectiveness. These classifiers have proven their mettle in predicting diabetes, swiftly categorizing new cases, and efficiently handling missing data. Their straightforward generation and understanding further reinforce their practicality [62].

Within the functions category, the SMO classifier has achieved a notable accuracy of 72.14%, showcasing its adaptability and efficiency. SMO, used for training support vector machines (SVM), simplifies the quadratic programming (QP) optimization problem into smaller, more manageable subproblems for efficient resolution. This technique not only reduces the memory needed for processing but also excels in managing large datasets. SMO includes preprocessing steps like replacing missing values, converting nominal data into binary forms, and normalizing data, all of which enhance prediction accuracy [63].

In the Bayes category, Bayes Nets have shown exceptional performance in various fields, such as aircraft systems, scientific research, and public safety [64]. However, they are not recommended for predictive tasks like those in this study because the algorithm focuses on assessing the impact of factors on outcomes, surpassing regression functions in determining variable impacts.

The KStar classifier had the lowest accuracy. This system handles noisy input well and requires a shorter training period. However, its performance significantly improves with larger datasets. Utilizing this method requires setting the parameter k's value, which entails substantial computational effort due to the need to calculate distances between training instances [65].

In conclusion, decision tree algorithms have achieved the highest accuracy levels and are recommended for use in classification and prediction projects. The accuracy of other algorithms is also competitive, suggesting their suitability for similar research. Additionally, integrating these algorithms with other deep learning or artificial intelligence strategies could enhance their effectiveness.

9.5 Summary

Researchers are keen to explore a variety of classifiers and develop new models to enhance the accuracy of diabetes prediction. This study shares that ambition, aiming to achieve high prediction accuracy. A detailed evaluation of machine learning (ML) and deep learning (DL) classifiers used over the past six years was performed. ML classifiers, particularly those used infrequently or not, were tested on the PID dataset to assess their effectiveness and provide insights into their application. These ML approaches achieved accuracy levels ranging from 68% to 74%. Meanwhile, DL algorithms reached a peak accuracy of 95%. Looking ahead, there is potential to employ these lesser-used classifiers on different datasets within a combined model, aiming to improve the accuracy of diabetes prediction further.

Acknowledgments

The authors would like to thank the management of BVRIT Hyderabad College of Engineering for Women, Hyderabad for providing the necessary facilities and resources to conduct to write this paper.

References

[1] Cho, N.; Shaw, J.; Karuranga, S.; Huang, Y.; Fernandes, J.D.R.; Ohlrogge, A.; Malanda, B. IDF Diabetes Atlas: Global estimates of diabetes prevalence for 2017 and projections for 2045. *Diabetes Res. Clin. Pr.* **2018**, *138*, 271–281.

[2] Sanz, J.A.; Galar, M.; Jurio, A.; Brugos, A.; Pagola, M.; Bustince, H. Medical diagnosis of cardiovascular diseases using an interval-valued fuzzy rule-based classification system. *Appl. Soft Comput.* **2014**, *20*, 103–111.

[3] Varma, K.V.; Rao, A.A.; Lakshmi, T.S.M.; Rao, P.N. A computational intelligence approach for a better diagnosis of diabetic patients. *Comput. Electr. Eng.* **2014**, *40*, 1758–1765.

[4] Kandhasamy, J.P.; Balamurali, S. Performance Analysis of Classifier Models to Predict Diabetes Mellitus. *Procedia Comput. Sci.* **2015**, *47*, 45–51.

[5] Iyer, A.; Jeyalatha, S.; Sumbaly, R. Diagnosis of Diabetes Using Classification Mining Techniques. *Int. J. Data Min. Knowl. Manag. Process.* **2015**, *5*, 1–14.

[6] Razavian, N.; Blecker, S.; Schmidt, A.M.; Smith-McLallen, A.; Nigam, S.; Sontag, D. Population-Level Prediction of Type 2 Diabetes from Claims Data and Analysis of Risk Factors. *Big Data* **2015**, *3*, 277–287.

[7] Ashiquzzaman, A.; Kawsar Tushar, A.; Rashedul Islam, M.D.; Shon, D.; Kichang, L.M.; Jeong-Ho, P.; Dong-Sun, L.; Jongmyon, K. Reduction of overfitting in diabetes prediction using deep learning neural network. In *IT Convergence and Security*; Lecture Notes in Electrical Engineering; Springer: Singapore, 2017; Volume 449.

[8] Swapna, G.; Soman, K.P.; Vinayakumar, R. Automated detection of diabetes using CNN and CNN-LSTM network and heart rate signals. *Procedia Comput. Sci.* **2018**, *132*, 1253–1262.

[9] Rahimloo, P.; Jafarian, A. Prediction of Diabetes by Using Artificial Neural Network, Logistic Regression Statistical Model and Combination of Them. *Bull. Société R. Sci. Liège* **2016**, *85*, 1148–1164.

[10] Gill, N.S.; Mittal, P. A computational hybrid model with two level classification using SVM and neural network for predicting the diabetes disease. *J. Theor. Appl. Inf. Technol.* **2016**, *87*, 1–10.

[11] NirmalaDevi, M.; Alias Balamurugan, S.A.; Swathi, U.V. An amalgam KNN to predict diabetes mellitus. In Proceedings of the 2013 IEEE International Conference ON Emerging Trends in Computing, Communication and Nanotechnology (ICECCN), Tirunelveli, India, 25–26 March 2013; pp. 691–695.

[12] Sun, Y.L.; Zhang, D.L. Machine Learning Techniques for Screening and Diagnosis of Diabetes: A Survey. *Teh. Vjesn.* **2019**, *26*, 872–880.

[13] Choudhury, A.; Gupta, D. A Survey on Medical Diagnosis of Diabetes Using Machine Learning Techniques. In *Recent Developments in Machine Learning and Data Analytics*; Springer: Singapore, 2019; pp. 67–68.
[14] Meherwar, F.; Maruf, P. Survey of Machine Learning Algorithms for Disease Diagnostic. *J. Intell. Learn. Syst. Appl.* **2017**, *9*, 1–16.
[15] Vijiyarani, S.; Sudha, S. Disease Prediction in Data Mining Technique—A Survey. *Int. J. Comput. Appl. Inf. Technol.* **2013**, *2*, 17–21.
[16] Deo, R.C. Machine Learning in Medicine. *Circulation* **2015**, *132*, 1920–1930.
[17] Yuvaraj, N.; SriPreethaa, K.R. Diabetes prediction in healthcare systems using machine learning algorithms on Hadoop cluster. *Clust. Comput.* **2017**, *22*, 1–9.
[18] Tafa, Z.; Pervetica, N.; Karahoda, B. An intelligent system for diabetes prediction. In Proceedings of the 2015 4th Mediterranean Conference on Embedded Computing (MECO), Budva, Montenegro, 14–18 June 2015; pp. 378–382.
[19] Sisodia, D.; Sisodia, D.S. Prediction of Diabetes using Classification Algorithms. *Procedia Comput. Sci.* **2018**, *132*, 1578–1585.
[20] Mercaldo, F.; Nardone, V.; Santone, A. Diabetes Mellitus Affected Patients Classification and Diagnosis through Machine Learning Techniques. *Procedia Comput. Sci.* **2017**, *112*, 2519–2528.
[21] Negi, A.; Jaiswal, V. A first attempt to develop a diabetes prediction method based on different global datasets. In Proceedings of the 2016 Fourth International Conference on Parallel, Distributed and Grid Computing (PDGC), Waknaghat, India, 22–24 December 2016; pp. 237–241.
[22] Olaniyi, E.O.; Adnan, K. Onset diabetes diagnosis using artificial neural network. *Int. J. Sci. Eng. Res.* **2014**, *5*, 754–759.
[23] Soltani, Z.; Jafarian, A. A New Artificial Neural Networks Approach for Diagnosing Diabetes Disease Type II. *Int. J. Adv. Comput. Sci. Appl.* **2016**, *7*, 89–94.
[24] Somnath, R.; Suvojit, M.; Sanket, B.; Riyanka, K.; Priti, G.; Sayantan, M.; Subhas, B. Prediction of Diabetes Type-II Using a Two-Class Neural Network. In Proceedings of the 2017 International Conference on Computational Intelligence, Communications, and Business Analytics, Kolkata, India, 24–25 March 2017; pp. 65–71.
[25] Mamuda, M.; Sathasivam, S. Predicting the survival of diabetes using neural network. In Proceedings of the AIP Conference Proceedings, Bydgoszcz, Poland, 9–11 May 2017; Volume 1870, pp. 40–46.

[26] Kumari, V.A.; Chitra, R. Classification of diabetes disease using support vector machine. *Int. J. Adv. Comput. Sci. Appl.* **2013**, *3*, 1797–1801.
[27] Farran, B.; Channanath, A.M.; Behbehani, K.; Thanaraj, T.A. Predictive models to assess risk of type 2 diabetes, hypertension and comorbidity: Machine-learning algorithms and validation using national health data from Kuwait—A cohort study. *BMJ Open* **2013**, *3*, 24–57.
[28] Tapak, L.; Mahjub, H.; Hamidi, O.; Poorolajal, J. Real-Data Comparison of Data Mining Methods in Prediction of Diabetes in Iran. *Healthc. Inform. Res.* **2013**, *19*, 177–185.
[29] Anand, R.; Kirar, V.P.S.; Burse, K. K-fold cross validation and classification accuracy of pima Indian diabetes data set using higher order neural network and PCA. *Int. J. Soft Comput. Eng.* **2013**, *2*, 2231–2307.
[30] Choi, S.B.; Kim, W.J.; Yoo, T.K.; Park, J.S.; Chung, J.W.; Lee, Y.H.; Kang, E.S.; Kim, D.W. Screening for Prediabetes Using Machine Learning Models. *Comput. Math. Methods Med.* **2014**, *2014*, 1–8.
[31] Sarwar, A.; Sharma, V. Comparative analysis of machine learning techniques in prognosis of type II diabetes. *AI Soc.* **2014**, *29*, 123–129.
[32] Durairaj, M.; Kalaiselvi, G. Prediction of Diabetes using Back propagation Algorithm. *Int. J. Innov. Technol.* **2015**, *1*, 21–25.
[33] Anand, A.; Shakti, D. Prediction of diabetes based on personal lifestyle indicators. In Proceedings of the 2015 1st International Conference on Next Generation Computing Technologies (NGCT), Dehradun, India, 4–5 September 2015; pp. 673–676.
[34] Malik, S.; Khadgawat, R.; Anand, S.; Gupta, S. Non-invasive detection of fasting blood glucose level via electrochemical measurement of saliva. *SpringerPlus* **2016**, *5*, 701.
[35] Perveen, S.; Shahbaz, M.; Guergachi, A.; Keshavjee, K. Performance Analysis of Data Mining Classification Techniques to Predict Diabetes. *Procedia Comput. Sci.* **2016**, *82*, 115–121.
[36] Joshi, S.; Borse, M. Detection and Prediction of Diabetes Mellitus Using Back-Propagation Neural Network. In Proceedings of the 2016 International Conference on Micro-Electronics and Telecommunication Engineering (ICMETE), Uttarpradesh, India, 22–23 September 2016; pp. 110–113.
[37] Sowjanya, K.; Singhal, A.; Choudhary, C. MobDBTest: A machine learning based system for predicting diabetes risk using mobile devices. In Proceedings of the 2015 IEEE International Advance Computing Conference (IACC), Bangalore, India, 12–13 June 2015; pp. 397–402.

[38] Cai, L.; Wu, H.; Li, D.; Zhou, K.; Zou, F. Type 2 Diabetes Biomarkers of Human Gut Microbiota Selected via Iterative Sure Independent Screening Method. *PLoS ONE* **2015**, *10*, e0140827.

[39] Maniruzzaman, M.; Kumar, N.; Menhazul Abedin, M.; Shaykhul Islam, M.; Suri, H.S.; El-Baz, A.S.; Suri, J.S. Comparative approaches for classification of diabetes mellitus data: Machine learning paradigm. *Comput. Methods Programs Biomed.* **2017**, *152*, 23–34.

[40] Mirshahvalad, R.; Zanjani, N.A. Diabetes prediction using ensemble perceptron algorithm. In Proceedings of the 2017 9th International Conference on Computational Intelligence and Communication Networks (CICN), Girne, Cyprus, 16–17 September 2017; pp. 190–194.

[41] Sun, X.; Yu, X.; Liu, J.; Wang, H. Glucose prediction for type 1 diabetes using KLMS algorithm. In Proceedings of the 2017 36th Chinese Control Conference (CCC), Liaoning, China, 26–28 July 2017; pp. 1124–1128.

[42] Mohebbi, A.; Aradóttir, T.B.; Johansen, A.R.; Bengtsson, H.; Fraccaro, M.; Mørup, M. A deep learning approach to adherence detection for type 2 diabetics. In Proceedings of the 2017 39th Annual International Conference of the IEEE Engineering in Medicine and Biology Society (EMBC), Jeju, Korea, 11–15 July 2017; pp. 2896–2899.

[43] Miotto, R.; Li, L.; Kidd, B.A.; Dudley, J.T. Deep Patient: An Unsupervised Representation to Predict the Future of Patients from the Electronic Health Records. *Sci. Rep.* **2016**, *6*, 26094.

[44] Pham, T.; Tran, T.; Phung, D.; Venkatesh, S. Predicting healthcare trajectories from medical records: A deep learning approach. *J. Biomed. Inform.* **2017**, *69*, 218–229.

[45] Balaji, H.; Iyengar, N.; Caytiles, R.D. Optimal Predictive analytics of Pima Diabetics using Deep Learning. *Int. J. Database Theory Appl.* **2017**, *10*, 47–62.

[46] Lekha, S.; Suchetha, M. Real-Time Non-Invasive Detection and Classification of Diabetes Using Modified Convolution Neural Network. *IEEE J. Biomed. Health Inform.* **2018**, *22*, 1630–1636.

[47] Askarzadeh, A.; Rezazadeh, A. Artificial neural network training using a new efficient optimization algorithm. *Appl. Soft Comput.* **2013**, *13*, 1206–1213.

[48] Rao, N.M.; Kannan, K.; Gao, X.Z.; Roy, D.S. Novel classifiers for intelligent disease diagnosis with multi-objective parameter evolution. *Comput. Electr. Eng.* **2018**, *67*, 483–496.

[49] Begg, R.; Kamruzzaman, J.; Sarkar, R. *Neural Networks in Healthcare: Potential and Challenges*; Idea Group Publishing: Hershey, PA, USA, 2006.

[50] Greeshma, U.; Annalakshmi, S. Artificial Neural Network (Research paper on basics of ANN). *Int. J. Sci. Eng. Res.* **2015**, 110–115.

[51] Zhang, G.; Patuwo, B.E.; Hu, M.Y. Forecasting with artificial neural networks: The state of the art. *Int. J. Forecast.* **1998**, *14*, 35–62.

[52] Srivastava, N.; Hinton, G.; Krizhevsky, A.; Sutskever, I.; Salakhutdinov, R. Dropout: A simple way to prevent neural networks from overfitting. *J. Mach. Learn. Res.* **2014**, *15*, 1929–1958.

[53] Vidyasagar, M. *Learning and Generalisation: With Applications to Neural Networks*; Springer Science & Business Media: London, UK, 2013.

[54] Maren, A.J.; Harston, C.T.; Pap, R.M. *Handbook of Neural Computing Applications*; Academic Press: Cambridge, MA, USA, 2014.

[55] Karayiannis, N.; Venetsanopoulos, A.N. *Artificial Neural Networks: Learning Algorithms, Performance Evaluation and Applications*; Springer Science & Business Media: London, UK, 2013; Volume 209.

[56] Goodfellow, I.; Bengio, Y.; Courville, A.; Bengio, Y. *Deep Learning*; MIT Press: Cambridge, MA, USA, 2016.

[57] Hassan, A. Deep Neural Language Model for Text Classification Based on Convolutional and Recurrent Neural Networks. Ph.D. Thesis, University of Bridgeport, Bridgeport, CT, USA, 2018.

[58] Kar, A.K. Bio inspired computing—A review of algorithms and scope of applications. *Expert Syst. Appl.* **2016**, *59*, 20–32.

[59] Naji, H.; Ashour, W. Text Classification for Arabic Words Using Rep-Tree. *Int. J. Comput. Sci. Inf. Technol.* **2016**, *8*, 101–108.

[60] Kumar, S.C.; Chowdary, E.D.; Venkat rama phani kumar, S.; Kishore, K.V.K. M5P model tree in predicting student performance: A case study. In Proceedings of the IEEE International Conference on Recent Trends in Electronics, Information & Communication Technology (RTEICT), Bangalore, India, 20–21 May 2016; pp. 1103–1107.

[61] Sharma, R.; Kumar, S.; Maheshwari, R. Comparative Analysis of Classification Techniques in Data Mining Using Different Datasets. *Int. J. Comput. Sci. Mobile Comput.* **2015**, *44*, 125–134.

[62] Fernández-Delgado, M.; Cernadas, E.; Barro, S.; Amorim, D. Do we need hundreds of classifiers to solve real world classification problems. *J. Mach. Learn. Res.* **2014**, *15*, 3133–3181.

[63] Platt, J. Fast Training of Support Vector Machines Using Sequential Minimal Optimization. In *Advances in Kernel Methods: Support Vector Learning, Advances in Kernel Methods—Support Vector Learning, Advances*; MIT Press: Cambridge, MA, USA, 1998; pp. 185–208. ISBN 0-262-19416-3.

[64] Su, J.; Zhang, H. Full Bayesian network classifiers. In Proceedings of the 23rd International Conference on Machine Learning, Pittsburgh, PA, USA, 25–29 June 2006; pp. 897–904.

[65] Mahmood, D.Y.; Hussein, M.A. Intrusion detection system based on K-star classifier and feature set reduction. *IOSR J. Comput. Eng.* **2013**, *15*, 107–112.

10

Intelligent Diagnosis Support System for Screening Diabetes Subjects using Hybrid Machine Learning Algorithms

Ch.Rajendra Prasad[1], Srinivas Samala[1], Sreedhar Kollem[1], Ravichander Janapati[1], Srikanth Yalabaka[1], and Moola Ramu[2]

[1]Department of ECE, SR University, India
[2]Department of ECE, Sumathi Reddy Institute of Technology for Women (SRITW), India
E-mail: chrprasad20@gmail.com; srinu486@gmail.com; ksreedhar829@gmail.com; ravichander.j@sru.edu.in; chinna7131@gmail.com; moola.ramu@gmail.com

Abstract

Diabetes mellitus is a serious health condition characterized by elevated blood sugar levels resulting from insufficient production of insulin and resistance to its effects. It is a leading cause of mortality worldwide. Recently, there has been increasing interest in employing artificial intelligence and machine learning techniques for the early detection of diabetes. This chapter presents a novel intelligent diagnosis support system designed to screen individuals for diabetes using machine learning. The study focuses on detecting diabetes mellitus by implementing machine learning algorithms: LightGBM and LightGBM plus KNN (Hybrid). The dataset utilized in this study is the Pima Indian Diabetes Database (PIDD) obtained from the National Institute of Diabetes and Digestive and Kidney Diseases (UCI ML Repository). The proposed machine learning algorithms are evaluated based on their area under the curve (AUC). The experimental results demonstrate that the LightGBM plus KNN outperforms in terms of accuracy, F1 score, precision, and recall in comparison with the LightGBM. Consequently, the developed intelligent system exhibits great potential in providing support for the early screening of diabetes subjects, thus aiding in timely intervention and management of the condition.

Keywords: Intelligent diagnosis, diabetes mellitus, machine learning, LightGBM, KNN, Pima Indian Diabetes Database.

10.1 Introduction

According to the 9th edition of the study published by the International Diabetes Federation (IDF), the global population of individuals between the ages of 20 and 64 who have been diagnosed or remain undiagnosed with diabetes stands at 351.7 million as of 2019. It is estimated that around 90% of these cases are classified as type-2 diabetes. According to a study, it is projected that the prevalence of diabetes would rise to 417.3 million individuals by the year 2030, and further escalate to 486.1 million individuals by 2045 [1]. The most significant surge is expected to occur in nations with low- and middle-income levels. As a result, diabetes mellitus (DM) has emerged as a critical worldwide health concern, necessitating timely identification and diagnosis in order to effectively mitigate and diminish its prevalence. Diabetes is a chronic health disorder characterized by impaired metabolic processes that hinder the body's capacity to convert food into usable energy. The process involves breaking down food into sugar and releasing it into the bloodstream, regulated by insulin [2]. In diabetes, the body either lacks sufficient insulin production or becomes resistant to its effects, leading to elevated blood sugar levels and potential health complications [3].

Type-1, type-2, and gestational diabetes are the three main kinds of diabetes and each has its own features. Type-1 diabetes results from an autoimmune reaction that impairs insulin production, typically diagnosed in children and young adults. Type-2 diabetes involves insulin resistance and is commonly diagnosed in adults, but increasingly in younger populations. Gestational diabetes occurs during pregnancy and may elevate the risk of health issues for both the mother and the baby. While there is no cure for diabetes, lifestyle changes such as weight management, healthy eating, and physical activity can greatly improve its management and prevention. More than 18% of pregnant women have type-III diabetes, according to a recent research [4]. Researchers in the medical area can benefit from data analysis because it allows them to extract useful information from datasets, allowing them to make well-informed decisions that advance the healthcare sector [5].

The immense significance of diabetes predicts the disease using machine learning (ML) techniques, an intriguing field of study. Decision-making forecasting based on pattern recognition, cluster analysis, and classification [6] are all data mining methodologies. For diabetes prediction, the present literature primarily employs supervised approaches. In this chapter, we looked at the state-of-the-art in applying ML techniques to forecast the onset of diabetes and other medical conditions. A research investigation was conducted utilizing self-organizing maps (SOM) and neural networks (NN), as well as principal component analysis (PCA), to predict diabetic analysis [7].

In another investigation, Kumar et al. developed a multimodal approach that utilized the support vector machine (SVM) method to pick features and remove unnecessary attributes from the dataset. For data clustering, a similar hybrid approach was developed utilizing Naive Bayes (NB) and K-means [8]. A similar investigation examined diabetes using the multilayer perceptron (MLP), NB, and decision tree algorithms. Using the PIMA dataset, they demonstrated that NB has superior performance to other approaches. In a similar vein, we use NN machine learning techniques to construct a model for heart disease [9, 10].

10.2 Related Work

Historically, healthcare professionals have predominantly utilized fixed data as a means to evaluate the management of diabetes. However, emerging studies indicate that the variability observed in these measurements could potentially offer equally or even more valuable insights. To test this hypothesis, extensive data from a group of individuals diagnosed with diabetes were analyzed, including their glycemic and lipid profiles over an extended period. Subsequently, the data was subjected to machine learning algorithms to analyze it and discern patterns of variability [11]. In [12], the authors employ a range of machine learning algorithms, such as logistic regression, support vector machines, decision trees, and neural networks, to develop prediction models. Accuracy, precision, and F1-score are all used as ways to measure performance. Early prediction of diabetes mellitus using machine learning models shows promise, according to the study's findings. The utilization of these models demonstrates their efficacy in accurately categorizing individuals who are susceptible to acquiring diabetes. Consequently, this enables healthcare practitioners to promptly implement appropriate therapies and lifestyle adjustments to avoid or effectively manage the condition.

In [13], the authors investigate the potential of machine learning techniques for forecasting diabetes mellitus onset and development. The paper explores various ML methods used to predict diabetes, such as SVM, decision trees, random forests, logistic regression, and artificial neural networks (ANN). The study's findings show that diabetes mellitus can be accurately predicted using machine learning methods applied to patient records. In terms of forecasting diabetes, various algorithms showed differing degrees of accuracy and robustness. This chapter examines the merits and disadvantages of several methods, emphasizing the significance of feature selection and hyperparameter tuning in maximizing the performance of models.

In [14], the authors investigate the use of machine learning algorithms for prenatal screening for gestational diabetes mellitus (GDM). If GDM is not diagnosed and treated in a timely manner, it can have serious consequences for both the mother and the baby. The authors have tackled this matter by utilizing machine learning techniques to construct a predictive model that can effectively detect women who are susceptible to developing gestational diabetes mellitus (GDM) during the initial 19 weeks of pregnancy. To build prediction models from the dataset, it was subjected to analysis using ML algorithms such as logistic regression, decision trees, and SVM. The objective of the models was to ascertain prospective cases of gestational diabetes mellitus (GDM) by utilizing the existing input factors. The findings of the study revealed encouraging prediction capacities of the machine learning models, demonstrating their promise in the early identification of gestational diabetes mellitus (GDM). The models demonstrated relatively good accuracy, sensitivity, and specificity factors that are essential for spotting at-risk people and reducing false positives.

A comprehensive discussion of data pre-processing techniques are essential, particularly in addressing missing values and implementing feature scaling. These procedures are crucial in ensuring the cleanliness and preparedness of the dataset for further analysis. The authors used machine learning techniques to create models that can predict who would get diabetes and how severe their disease will be. To evaluate the effectiveness of each algorithm's ability to predict diabetes, its performance is measured by employing a variety of measures, including F1-score, accuracy, recall, and precision [15]. The application of artificial intelligence (AI) methodologies, particularly ML, in the provision of ehealthcare services and support for persons diagnosed with type-2 diabetes mellitus (T2DM), was presented in [16].

The utilization of machine learning techniques in several domains of T2DM care, including risk assessment, early detection, management of glucose levels, and customization of treatment approaches are also discussed. The chapter elucidates the obstacles and constraints entailed in the implementation of AI in the domain of diabetes care. These challenges encompass issues pertaining to data quality, interpretability of models, and regulatory factors that warrant consideration. It emphasizes the significance of conducting extensive validation and clinical studies to guarantee the security and efficacy of AI-based interventions.

A wide range of supervised and unsupervised methods such as k-means clustering, neural networks, and decision tree algorithms were presented in [17]. The experimental outcomes show that the machine learning methods are efficient in dividing diabetes mellitus patients into distinct subgroups, such as those with type-1, type-2, or gestational diabetes. Furthermore, the algorithms show promise for predicting disease progression and consequences, which could inform more targeted therapeutic approaches. Several ML models (random forest, decision tree, gradient boosting, and generalized linear model) are trained and evaluated using data that includes clinical, demographic, and lifestyle aspects [18]. These models are utilized to construct predictive tools that can measure an individual's chances of emergent type-2 diabetes. In this study, the proposed algorithms are put through rigorous testing and comparison to see how accurately, sensitively, and specifically they can detect diabetes in its early stages. The results of the study demonstrate encouraging outcomes, suggesting that prediction models utilizing machine learning techniques can proficiently discern individuals who are susceptible to type-2 diabetes.

To aid in the early detection of diabetes mellitus (DM), a novel method based on ML algorithms is proposed in [19]. The authors make use of a systematic methodology and the diabetes in Pima Indians dataset. The dataset is trained and evaluated using six distinct ML algorithms: Naive Bayes, random forest, k-nearest neighbors (KNN), logistic regression, and decision tree. Different measures are employed to examine the effectiveness of these algorithms. The experiential findings suggest that the XGBoost method attains the maximum level of accuracy, with a score of 88.2%, hence exceeding alternative machine learning models. Because of this, XGBoost appears to be a useful technique for detecting DM in its earliest stages.

Type-1 diabetes mellitus (T1DM) can use big data and machine learning methods to forecast their blood sugar levels in the near future [20]. To avoid health problems and consequences, people with T1D must keep their blood glucose levels under control. Data on CGM readings, insulin doses, food intake, and exercise patterns were obtained from a large sample of people with type-1 diabetes. To examine this dataset, they used several machine learning strategies like regression models, decision trees, and deep neural networks. The objective was to develop reliable forecasting algorithms for predicting blood glucose levels across brief time horizons. The study's findings suggest that machine learning approaches applied to vast amounts of data from CGMs have the potential to improve short-term blood glucose prediction. These predictive models have proven to be highly accurate and reliable, providing insights that can help people with T1DM better manage their illness.

10.3 Materials and Methods

The proposed model architecture is illustrated in Figure 10.1. In this architecture, the raw data is collected from the Pima Indians Diabetes Database [21]. The collected data is pre-processed using median and min−max normalization methods. Then, ML models such as light gradient boosting machine (LightGBM) and LightGBM plus K-nearest neighbor models are used to predict diabetes subjects. Further, the comprehensive description of the model is presented in the following subsections.

10.3.1 Dataset

The dataset utilized in this investigation was obtained from the Kaggle website [22], specifically the Pima Indians Diabetes Database. This dataset is originally from the National Institute of Diabetes and Digestive and Kidney Diseases. Table 10.1 shows the Pima Indians Diabetes Database parameters.

Table 10.2 illustrates the eight features of the dataset, along with the output class (0 or 1), which serves as a binary indicator of the presence or absence of diabetes.

10.3.2 Data pre-processing

It is common practice in data analysis studies to substitute the mean of the related feature for missing values (for instance, the column mean in tabular data). However, for small datasets, using the median of the related

Table 10.1 Dataset parameters.

Patients	Subjects	Features	Classes
Women 21+ of Pima Indian descent	768	8	2

10.3 Materials and Methods 229

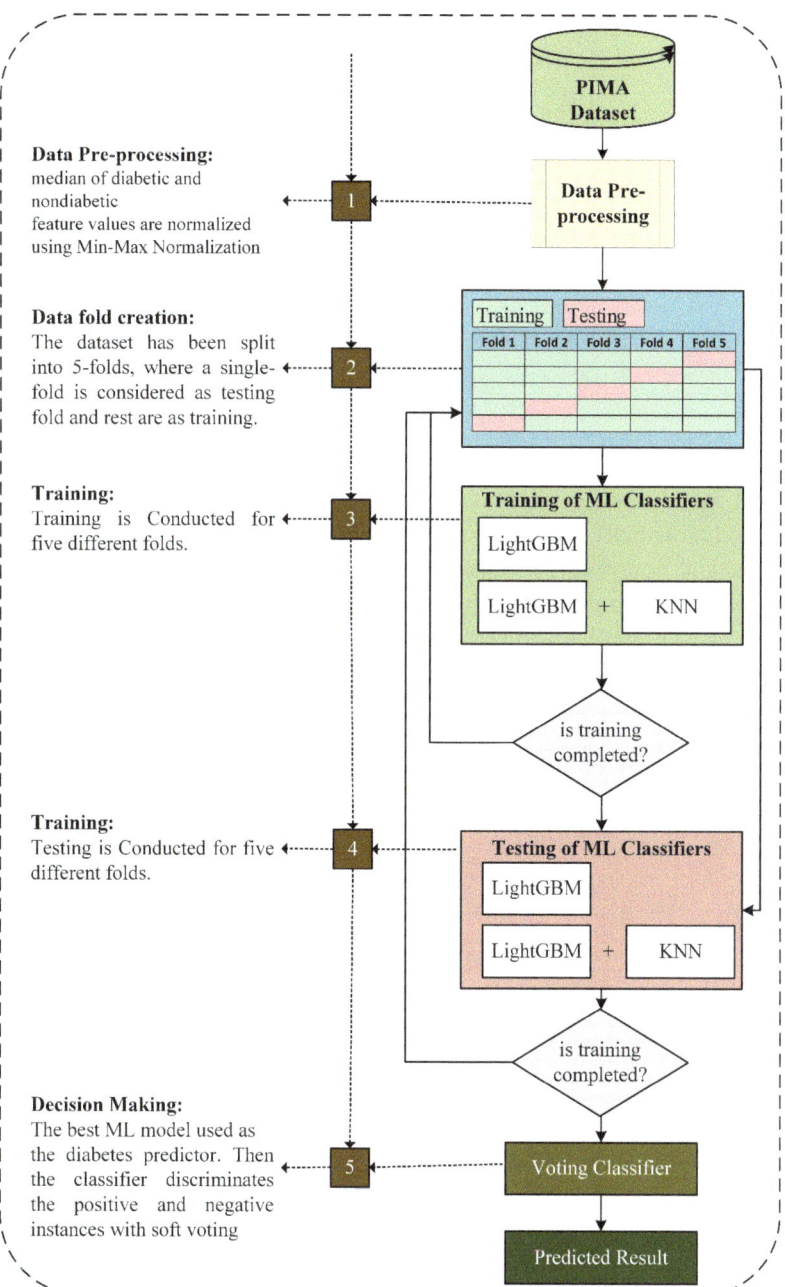

Figure 10.1 Proposed model architecture.

Table 10.2 Output classes and features of the dataset.

Parameters	Std deviation	Mean	Quantiles				
			Max	75%	50%	25%	Min
Age	11.8	33.2	81	41	29	24	21
Pregnancies	3.37	3.85	17	6	3	1	0
Blood pressure	19.3	69.1	122	80	72	62	0
Glucose	32	121	199	141	117	99	0
Insulin	115	79.8	846	128	32	0	0
Skin thickness	15.9	20.5	99	32	23	0	0
Diabetes pedigree function	0.33	0.47	2.42	0.63	0.37	0.24	0.08
Body mass index	7.88	32	67.1	36.6	32	27.3	0
Outcome	0.48	0.35	1	1	0	0	0

characteristics rather than the mean to replace missing values is preferable [22]. To fill in the blanks, we utilized a tried-and-true method: we took the feature's median value.

To make the new figures more accurate, we calculated the median for both diabetes and non-diabetic patients independently. When developing a machine learning model, it is important to keep in mind that the feature values span a variety of intervals, which may impact the final product. To ensure that all features have values between 0 and 1, we use the min−max normalization approach [23, 24] to normalization approach [25, 26] to normalize the feature values. Here, we assume an initial set of features $\mathbf{X} = \{x_0, x_1, \ldots, x_{n-1}\}$ with n entries; the average is

$$x'_i = \left[\frac{(x_i - \min(X))}{\max(X) - \min(X)} \right],$$

where $i \in \{0, 1, 2, \ldots, n-1\}$, the minimal value of $\mathbf{X}$ is represented by $\min(\mathbf{X})$, and the largest value by $\max(\mathbf{X})$, and $\mathbf{X}' = \{x'0, x'1, \ldots, x'_{n-1}\}$ features that have been normalized to $0 \leq x'_i \leq 1$

10.3.3 Machine learning models

Several numbers of classifiers were presented in the literature. The LightGBM machine learning classifier is employed in the proposed model due to its advantages. LightGBM's strengths lie in its capacity to process massive amounts of data without slowing down during training, as well as in its support for parallel and GPU-based learning and its reduced memory and CPU requirements.

LightGBM:

Unlike many boosting algorithms, gradient boosting decision tree (GBDT) relies little on feature selection because it employs cooperative weak decision tree classifiers [11]. The basis for GBDT is decision trees, where the estimates from individual tree in the chain are combined together to determine the ultimate choice. At each stage, a new decision tree is generated to account for the discrepancy between the most recent forecast and the actual data. GBDT was chosen by many scientists because of its reliability, effectiveness, and interpretability [12]. During a training session **TR = {(tr1 1, tr2 1, ..., trr 1), (tr1 2, tr2 2, ..., trr 2), ..., (tr1 m, tr2 m, ..., trr m)}** of size **m × r** dimensions, where tr represents training examples and trr represents class labels, and f(TR) represents the expected and optimized result. Assuming this, we can estimate f(TR) so as to minimize **L(trr, f(TR))**.

$$\hat{f} = \underset{f}{\operatorname{argmin}}\, E_{\mathrm{TR},\mathrm{tr}^r}\left[L\left(\mathrm{tr}^r, f(\mathrm{TR})\right)\right],$$

in addition to determining GBDT iteration criteria.

$$F_k(\mathrm{TR}) =$$

$$F_{k-1}(\mathrm{TR}) + h_k(\mathrm{TR})\left[\underset{f}{\operatorname{argmin}} \sum_{i=1} L\left((\mathrm{tr}^r, F_{k-1}(\mathrm{TR}_i) + \gamma h_k(\mathrm{TR}_i))\right)\right].$$

In this case, **k** is the iterations, and $\mathbf{h}_k(\mathbf{TR})$ is the first decision tree from the training data. While the GBDT classification method performs well on modest datasets, it falters under the weight of a growing number of high-dimensional datasets [13].

KNN:

K-nearest neighbor, or KNN for short, is a classification technique for supervised learning [27, 28]. It is a straightforward method for sorting new instances into groups based on their similarities. In KNN, we can use a variety of methods to determine which of two data points is closest to one another. However, KNN requires additional memory due to the fact that all of the training data must be kept in-memory. Despite this, KNN is widely employed in medical diagnostics, with a particular emphasis on the diagnosis of diabetes. Multiple distance-measuring methods are compatible with the KNN. In this example, though, we are taking into account the Euclidean distance. Each **TR** training sample has been ordered by increasing distance values. The row with the shortest distance should have its predicted label for the relevant training instance match the class label in that row. On the other hand, in accordance with KNN, the **k** number of rows that have small distance to the target instance are the ones that are selected. The target instance's projected classes are the classes assigned to the k most similar instances. The KNN algorithm and its underlying working concept given as pseudocode are as follows:

KNN algorithm pseudocode
Inputs
Training Instances(TR)$_{m \times r}$ = $\{(tr_1^1, tr_1^2, ..., tr_1^r), (tr_2^1, tr_2^2, ..., tr_2^r), ..., (tr_m^1, tr_m^2, ..., tr_m^r)\}$ *Testing Instances(TS)$_{n \times r}$ =* $\{(tr_1^1, tr_1^2, ..., tr_1^r), (tr_2^1, tr_2^2, ..., tr_2^r), ..., (tr_n^1, tr_n^2, ..., tr_n^r)\}$ *K= Number of neighbors*
Process

begin
    for i := 1 to n
       for j := 1 to n
          D[j]=Euclidean(ts$_i$, tr$_j$)
    end
    SI = Sort(TR |D)$_{asc}^{D}$
    SSI = k(SI)
    rt_n^r = majority_vote(SSI$_{0\,to\,k}^r$)
    end
    return **TS**
 end

Output
Attribute t_n^r of TS populated with projected class labels.

Prediction of diabetes with the proposed models:

The prediction of diabetes with the proposed models is employed in four stages such as data folding, training, testing, and decision making as illustrated in Figure 10.1. In stage 1, the data is set up into five sections. For each of the five blocks, five iterations of training and testing have been performed on the ensembles, with one data fold employed for testing and the other four folds used for training. During stage 2, the classifiers LightGBM and LightGBM + KNN are prepared. The classifiers are subsequently trained using a fourfold cross-validation approach and evaluated using a single fold of data. The results of five training and testing cycles have been realized. The prediction of the suggested machine learning models is obtained in the final stage through the utilization of a soft voting mechanism. The soft voting for the suggested machine learning models is accomplished by the following process

$$\hat{y} = \underset{i}{\operatorname{argmix}} \sum_{j=1}^{k} w_j p_{ij}.$$

The weight allocated to the **jth** classifier is denoted as $\mathbf{w}_j$, whereas $\mathbf{p}$ represents the predicted probabilities.

10.4 Results and Discussion

The experimentation is conducted utilizing a laptop equipped with an Intel Core i5 7200U CPU, a dedicated graphics processing unit featuring a Geforce 940 MX, and 8 GB of random-access memory. To evaluate the performance of the proposed model, the following performance metrics are considered:

$$Accuracy = \frac{Number\ of\ correct\ predictions}{Total\ Number\ of\ predictions}$$

$$Precision = \frac{Trpos}{Trpos + Flspos}$$

$$Recall = \frac{Trpos}{Trpos + Flsneg}$$

$$F1\ Score = 2 \times \left[\frac{Recall \times Precision}{Recall + Precision} \right]$$

The proposed ML model incorporates a voting classifier for implementation [29, 30]. These models are sometimes referred to as meta-algorithms due to their ability to integrate many machine learning approaches into a

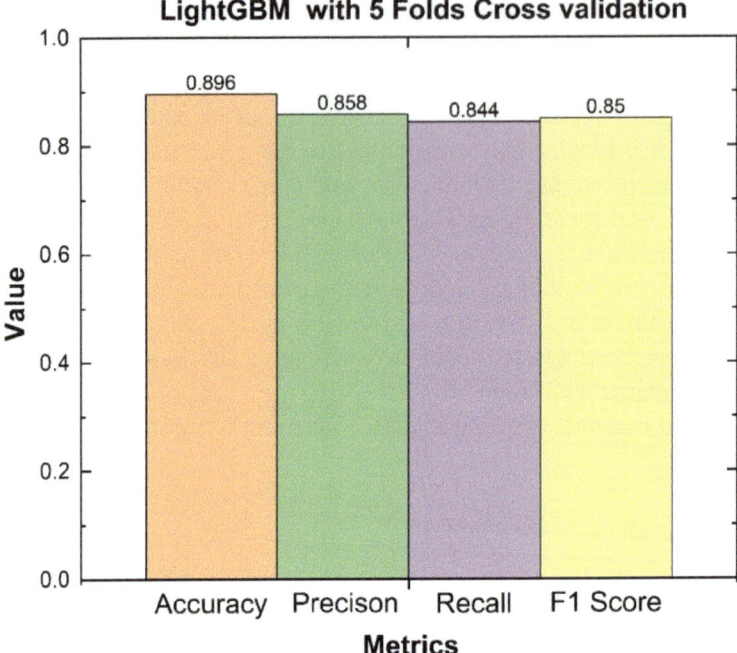

Figure 10.2 Mean performance metrics with fold cross-validation of the LightGBM.

Table 10.3 LightGBM with fivefold cross-validation.

Fold	Accuracy	Precision	Recall	F1 score	ROC AUC
1	0.903	0.915	0.796	0.851	0.945
2	0.864	0.789	0.833	0.811	0.926
3	0.896	0.865	0.833	0.849	0.949
4	0.889	0.846	0.83	0.838	0.944
5	0.928	0.875	0.925	0.899	0.972
Mean	0.896	0.858	0.844	0.85	0.947
Standard deviation	0.021	0.041	0.043	0.029	0.015

single prediction model. In this particular scenario, the classifier methods of LightGBM and LightGBM + KNN are employed as classifiers with the incorporation of hyperparameter adjustment. The soft voting approach is employed for decision-making purposes. The performance metrics of the ensemble model are assessed using a fivefold cross-validation approach for each implementation. Table 10.3 presents the outcomes of fivefold cross-validation for several LightGBM models. The mean performance metrics with fold cross-validation are illustrated in Figure 10.2.

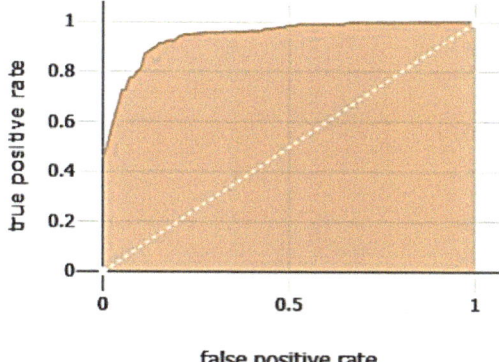

Figure 10.3 ROC curve of the LightGBM.

Figure 10.3 illustrates the construction of the receiver operating characteristic (ROC) curve for LightGBM, which involves graphing the true positive rate (TPR) against the false positive rate (FPR) at different threshold values. The ROC curve value for LightGBM is about 0.943.

Figure 10.4 depicts the accuracy–recall curve, which showcases the tradeoff between accuracy and recall at various thresholds in the context of LightGBM.

As a single classifier, the LightGBM may not reach an accuracy of 90%. To achieve better accuracy, we implemented a hybrid classifier using LightGBM and KNN. The Table 10.4 represents the LightGBM and KNN with fivefold cross-validation results metrics. The experimental evaluation of

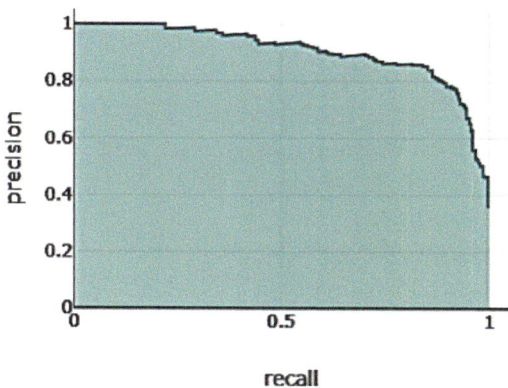

Figure 10.4 Precision–recall curve of the LightGBM.

Table 10.4 LightGBM and KNN with fivefold cross-validation.

Fold	Accuracy	Precision	Recall	F1 score	ROC AUC
1	0.896	0.896	0.796	0.843	0.922
2	0.877	0.797	0.87	0.832	0.918
3	0.916	0.902	0.852	0.876	0.937
4	0.902	0.88	0.83	0.854	0.94
5	0.941	0.893	0.943	0.917	0.953
Mean	0.906	0.873	0.858	0.865	0.934
Standard deviation	0.021	0.039	0.049	0.03	0.013

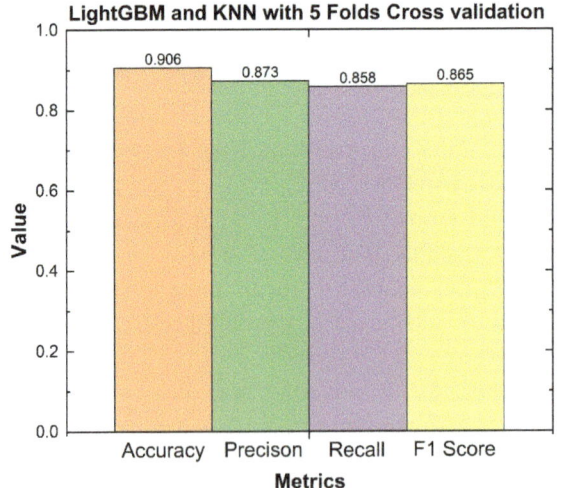

Figure 10.5 Mean performance metrics with fold cross-validation of the LightGBM + KNN.

the performance metrics of the LightGBM + KNN model is conducted with respect to the number of folds utilized in the cross-validation method. With the exception of the fivefold cross-validated scores, all other cross-validated data demonstrated a little decrease in performance metrics. The selection of fivefold cross-validation is motivated by its ability to mitigate the issues of time constraints and over-fitting, as well as its demonstrated superiority in terms of total performance scores and computational efficiency.

The mean performance metrics with fold cross-validation are illustrated in Figure 10.5. The main advantage of this hybrid machine learning classifier is a computationally efficient model compared to the LightGBM model. Therefore, the LightGBM + KNN is the better choice for the prediction of diabetes subjects.

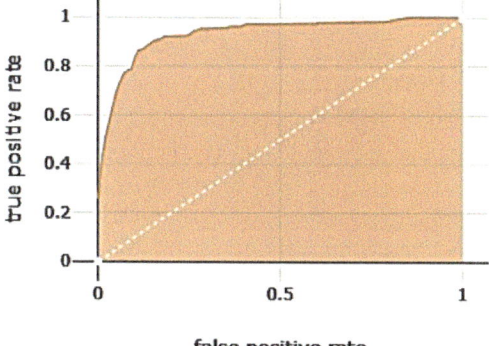

Figure 10.6 ROC curve of the LightGBM.

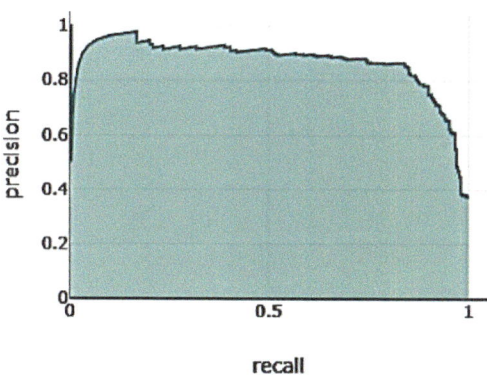

Figure 10.7 Precision−recall curve of the LightGBM.

The ROC curve of the LightGBM + KNN is illustrated in Figure 10.6. The ROC curve value for LightGBM is about 0.933. The precision−recall curve of the LightGBM + KNN model is illustrated in Figure 10.7. When prioritizing performance, the ideal method would be to utilize LightGBM in combination with KNN due to the decreased number of classifiers.

10.5 Conclusion

This chapter presented an intelligent diagnosis support system for screening diabetes subjects using LightGBM and LightGBM plus KNN algorithms. The evaluation metrics such as accuracy, precision, recall, F1-score, and ROC are analyzed. The model's performance was evaluated for fivefold cross-validation to yield the best results. The fivefold cross-validated mean accuracies of LightGBM and LightGBM plus KNN algorithms are 0.896 and 0.906, respectively. However, the ROC curve is better in LightGBM as a comparison with the LightGBM plus KNN. The LightGBM plus KNN is computationally efficient due its lower number of classifiers. The results of this study reveal that the hybrid machine learning model is suitable for the prediction of diabetic subjects.

References

[1] Rufo, D. D., Debelee, T. G., Ibenthal, A., & Negera, W. G. (2021). Diagnosis of diabetes mellitus using gradient boosting machine (LightGBM). Diagnostics, 11(9), 1714.
[2] S. Siddiqui, Depression in type 2 diabetes mellitus—a brief review. Diabetes Metab. Synd.Clin. Res. Rev. 8(1), 62–65 (2014)
[3] K. Rajesh, V. Sangeetha, Application of data mining methods and techniques for diabetes diagnosis. Int. J. Eng. Innov. Technol. (IJEIT) 2(3) (2012)
[4] S. Sarma Kattamuri, Predictive modeling with SAS enterprise miner: practical solutions for business applications (SAS Institute, 2013)
[5] I. Yoo et al., Data mining in healthcare and biomedicine: a survey of the literature. J. Med. Syst. 36(4), 2431–2448 (2012)
[6] Schultz, Matthew G., et al. "Data mining methods for detection of new malicious executables." Proceedings 2001 IEEE Symposium on Security and Privacy. S&P 2001. IEEE, 2000.
[7] MehrbakhshNilashia Accuracy Improvement for Diabetes Disease Classification: A Case on a Public Medical Dataset", Fuzzy Information and Engineering, Volume 9, Issue 3, September 2017, Pages 345-357.
[8] Kumar, Binit, et al. "Retinal neuroprotective effects of quercetin in streptozotocin-induced diabetic rats." Experimental Eye Research 125 (2014): 193-202.
[9] Pandeeswari, L., Rajeswari, K.: K-means clustering and Naïve Bayes classifier for categorization of diabetes patients. Int. J. Innov. Sci. Eng. Technol. (IJISET) 2(1) (2015)
[10] Koklu, M., Unal, Y.: Analysis of a population of diabetic patients databases with classifiers. World Acad. Sci. Eng. Technol. 7(8) (2013)
[11] Lee, S., Zhou, J., Wong, W. T., Liu, T., Wu, W. K., Wong, I. C. K., ... & Tse, G. (2021). Glycemic and lipid variability for predicting

complications and mortality in diabetes mellitus using machine learning. BMC endocrine disorders, 21, 1-15.
[12] Tripathi, G., & Kumar, R. (2020, June). Early prediction of diabetes mellitus using machine learning. In 2020 8th international conference on reliability, Infocom technologies and optimization (trends and future directions)(ICRITO) (pp. 1009-1014). IEEE.
[13] Lai, H., Huang, H., Keshavjee, K., Guergachi, A., & Gao, X. (2019). Predictive models for diabetes mellitus using machine learning techniques. BMC endocrine disorders, 19, 1-9.
[14] Xiong, Y., Lin, L., Chen, Y., Salerno, S., Li, Y., Zeng, X., & Li, H. (2022). Prediction of gestational diabetes mellitus in the first 19 weeks of pregnancy using machine learning techniques. The journal of maternal-fetal & neonatal medicine, 35(13), 2457-2463.
[15] Alehegn, M., Joshi, R., & Mulay, P. (2018). Analysis and prediction of diabetes mellitus using machine learning algorithm. International Journal of Pure and Applied Mathematics, 118(9), 871-878.
[16] Abhari, S., Kalhori, S. R. N., Ebrahimi, M., Hasannejadasl, H., & Garavand, A. (2019). Artificial intelligence applications in type 2 diabetes mellitus care: focus on machine learning methods. Healthcare informatics research, 25(4), 248-261.
[17] Mercaldo, F., Nardone, V., & Santone, A. (2017). Diabetes mellitus affected patients classification and diagnosis through machine learning techniques. Procedia computer science, 112, 2519-2528.
[18] Kopitar, L., Kocbek, P., Cilar, L., Sheikh, A., & Stiglic, G. (2020). Early detection of type 2 diabetes mellitus using machine learning-based prediction models. Scientific reports, 10(1), 11981.
[19] Miriyala, N. P., Kottapalli, R. L., Miriyala, G. P., Lorenzini, G., Ganteda, C., & Bhogapurapu, V. A. (2022). Diagnostic analysis of diabetes mellitus using machine learning approach. Revue d'Intelligence Artificielle, 36(3), 347.
[20] Rodríguez-Rodríguez, I., Chatzigiannakis, I., Rodríguez, J. V., Maranghi, M., Gentili, M., & Zamora-Izquierdo, M. Á. (2019). Utility of big data in predicting short-term blood glucose levels in type 1 diabetes mellitus through machine learning techniques. Sensors, 19(20), 4482.
[21] https://www.kaggle.com/datasets/uciml/pima-indians-diabetes-database
[22] Xu, Z.; Wang, Z. A risk prediction model for type 2 diabetes based on weighted feature selection of random forest and xgboost ensemble

classifier. In Proceedings of the 2019 Eleventh International Conference on Advanced Computational Intelligence (ICACI), Guilin, China, 7–9 June 2019; pp. 278–283.
[23] Al Shalabi, L.; Shaaban, Z. Normalization as a preprocessing engine for data mining and the approach of preference matrix. In Proceedings of the 2006 International Conference on Dependability of Computer Systems, Szklarska Poreba, Poland, 25–27 May 2006; pp. 207–214.
[24] Song, W., et al.: Design of a flexible wearable smart sEMG recorder integrated gradient boosting decision tree based hand gesture recognition. IEEE Trans. Biomed. Circuits Syst. 13(6), 1563–1574 (2019)
[25] Zhang, Z., Jung, C.: GBDT-MO: Gradient-Boosted Decision Trees for Multiple Outputs. IEEE Trans. Neural Netw. Learn. Syst. 32(7), 3156–67 (2020)
[26] Chen, C., Zhang, Q., Ma, Q., Yu, B.: LightGBM-PPI: predicting protein-protein interactions through LightGBM with multiinformation fusion. Chemom. Intell. Lab. Syst. 191, 54–64 (2019)
[27] Srivastava, T., Srivastava, T.: Introduction to k-NN, k-nearest neighbors: Simplifed. Anal. Vidhya (2014)
[28] Zhang, Z.: Introduction to machine learning: k-nearest neighbors. Ann. Transl. Med. 4(11) (2016)
[29] Kittler, J., Hatef, M., Duin, R.P.W., Matas, J.: On combining classifers. IEEE Trans. Pattern Anal. Mach. Intell. 20(3), 226–239 (1998)
[30] Kuncheva, L.I.: Combining Pattern Classifers: Methods and Algorithms, 2nd edn. Wiley, Hoboken, NJ, USA (2014)

11

Cyber–Physical System for Managing Diabetic Healthcare

Usha Desai[1], Kandala N. V. P. S. Rajesh[2], T. Kishore Kumar[3], Varadraj Gurupur[4], and Ganesh R. Naik[5]

[1]Department of Electronics and Communication Engineering, S.E.A College of Engineering & Technology, Bengaluru, India
[2]School of Electronics Engineering, VIT-AP University, India
[3]Department of Electronics and Communication Engineering, NIT Warangal, India
[4]School of Global Health Management and Informatics, University of Central Florida, USA
[5]College of Medicine and Public Health, Flinders University, South Australia
E-mail: dr.ushadesai@seaedu.ac.in; kandala.rajesh2014@gmail.com; kishoret@nitw.ac.in; varadraj.gurupur@ucf.edu; ganesh.naik@flinders.edu.au

Abstract

Managing diabetes effectively can be challenging for caretakers who are responsible for monitoring glucose levels, administering medications, and making informed decisions about diet and lifestyle choices. The complexity of diabetes management and the need for personalized care requires a comprehensive decision support system to assist caretakers in making informed decisions and providing optimal care to individuals with diabetes. The decision support system (DSS) for diabetes caretakers has the potential to revolutionize diabetes management by leveraging data-driven insights, personalized recommendations, and enhanced caretaker–patient collaboration. It can lead to improved glucose control, reduced complications, and enhanced quality of life for individuals with diabetes. By empowering caretakers with actionable information, the system can alleviate the burden of diabetes management, enhance caretaker confidence, and ultimately contribute to better patient outcomes. Furthermore, the system can provide valuable data for research and population health management, enabling healthcare providers and policymakers to make informed decisions and develop targeted interventions to address the growing burden of diabetes on a broader scale.

Keywords: Decision support system (DSS), diabetic healthcare, wearable device, AI-based diabetic management.

11.1 Background

Diabetes mellitus (DM) is a metabolic disorder described by high blood glucose levels. The two main subtypes of DM are type-1 diabetes mellitus (T1DM) and type-2 diabetes mellitus (T2DM), each having distinct causes and mechanisms. T1DM typically appears in children or adolescents and arises from inadequate insulin secretion. Besides, T2DM is commonly associated with middle-aged and older adults who have sustained high blood sugar levels due to unhealthy lifestyle choices and dietary habits. The pathogenesis of T1DM and T2DM varies significantly and requires different treatment approaches for each type [1].

11.1.1 Epidemiology of DM in India

According to the World Health Organization (WHO) [2], in India, approximately 77 million adults aged 18 and above have T2DM, while nearly 25 million individuals are considered prediabetic, putting them at a higher risk of developing diabetes soon. Alarmingly, over 50% of people are unaware of their diabetic condition, leading to potential health complications if left undetected and untreated. Adults with diabetes face a significantly increased risk of heart attacks and strokes, with a two to three times higher likelihood. Additionally, reduced blood flow and nerve damage in the feet, known as neuropathy, elevate the chances of foot ulcers, infections, and the need for limb amputation. Diabetic retinopathy, resulting from long-term damage to the small blood vessels in the retina, is a significant cause of blindness. Moreover, diabetes ranks among the primary causes of kidney failure.

Primary facts [3]: (i) In 2019 alone, the deaths due to DM and kidney diseases because of DM are approximately 2 million. (ii) The rise of the DM is 108 million to 422 million from 1980 to 2014, approximately a 300% increase within 34 years. On average, a 100% increment is observed for a decade.

11.1.2 Epidemiology of DM in USA

According to the National Institute of Diabetes and Digestive and Kidney Diseases (NIDDK) [4], around 37 million people of age 18 and above (~11% of the US population) are suffering from T2DM, which is almost 98% of DM of all ages in 2019. Among them, 28.5 million people are diagnosed with DM, and the rest are undiagnosed (about 25% of subjects with DM).

Primary facts: (i) Among all the adult (aged 18 and above) US population, almost one in three had prediabetes in 2019. (ii) Gender-wise, more men (41%) than women (32%) had the above-mentioned condition between 2017 and 2020. (iii) Especially, in the age group of 12−18 years, one out of five persons had prediabetes between 2017 and 2020. Recent years have witnessed a surge in the popularity and advancements of continuous glucose monitoring (CGM) systems, which allow for real-time monitoring of glucose levels and offer valuable data for effective diabetes management. However, the effectiveness of these technologies remains inconclusive and evidence is scattered. Therefore, conducting a systematic examination of these technologies is crucial for informing the development of future advancements in this field. In this review, we explore the latest research and developments on wearable glucose monitoring devices for diabetes management. The review draws upon scientific literature from PubMed, utilizing the keyword "wearable glucose monitoring devices for diabetes management." A search conducted on PubMed using the keyword "wearable glucose monitoring devices for diabetes management" yielded a total of 103 articles spanning a period of 10 years (2014−2023). These articles include various types of publications such as books and documents, clinical trials, meta-analyses, randomized controlled trials, and systematic reviews. The distribution of these articles across different years is depicted in Figure 11.1. To obtain a comprehensive overview of the objective from the pool of articles, a filter option was utilized to focus specifically on systematic review articles. Systematic reviews are known to provide detailed insights into wearable glucose monitoring devices for diabetes management over the years. This filter yielded a total of six review articles published between 2017 and 2023. The distribution of these review articles across different years is depicted in Figure 11.2. This statement can be removed and it is adding nothing here. Wang, Youfa, et al. [8] conducted a study that analyzed research articles on mobile health (mHealth) related to obesity and diabetes. The focus of the study was on the period between 2000 and 2016, using the PubMed database. The researchers included 24 studies in their analyses, considering factors such as sample size, ethnicity, and gender. The mHealth studies were categorized into three types: mobile

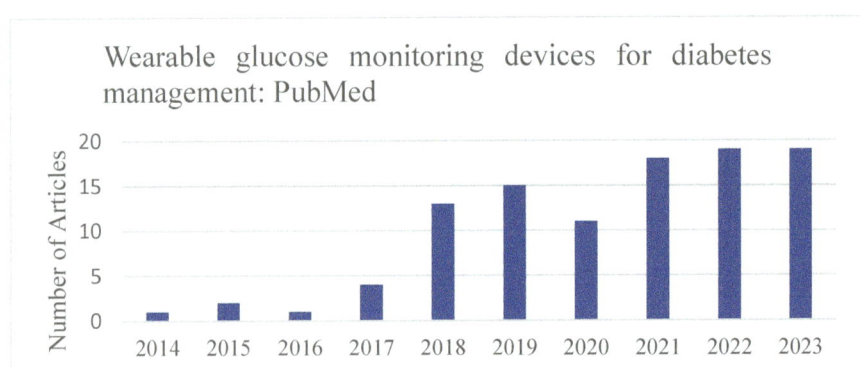

Figure 11.1 Number of articles available in PubMed with the keyword "Wearable glucose monitoring devices for diabetes management."

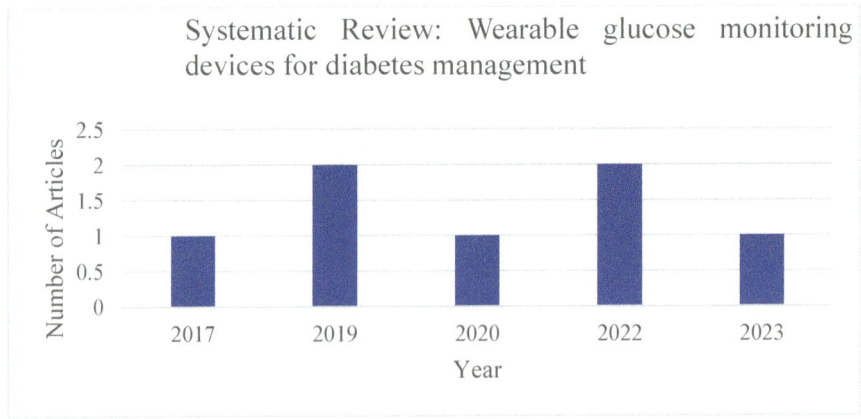

Figure 11.2 Number of systematic review articles on wearable glucose monitoring devices for diabetes management from 2017 to 2023.

text message interventions, wearable device interventions, and smartphone application interventions.

The findings of the study revealed several interesting facts. Among the 24 mHealth intervention studies, 54% (13) utilized mobile text messages, 25% (6) used wearable devices, and 21% (5) employed mobile applications. Regarding the duration of interventions, only a small number of studies (4, 17%) had interventions lasting longer than six months but less than two years.

The researchers observed that mobile applications and text messages were primarily used to provide tips and knowledge for weight control and blood glucose level management. These interventions also aimed to offer regular reminder services. Additionally, the mobile applications collected physiological information from patients to assist in self-monitoring and disease management.

In contrast, wearable devices were used to gather necessary data from patients, which were then integrated into mobile platforms for continuous monitoring. These interventions yielded more accurate results but involved greater complexity. Furthermore, these studies were more focused on long-term interventions. The positive outcomes and responses observed in patients' behavioral changes, such as reductions in obesity and improved diabetes control, served as a motivating factor in this research.

11.1.3 Limitations and future directions

However, most studies included small samples and short intervention periods and did not use rigorous data collection or analytic approaches. Although some studies suggest that mHealth interventions are effective and promising, most are pilot studies or have limitations in their study designs. There is an essential need for future studies that use larger study samples, longer intervention (>6 months) and follow-up periods (>6 months), and integrative and personalized innovative mobile technologies to provide comprehensive and sustainable support for patients and health service providers.

11.1.4 Methods to prevent or reduce the effect of DM

Adopting a healthy diet, engaging in regular physical activity, maintaining normal body weight, and abstaining from tobacco use are effective measures to prevent or delay the onset of type-2 diabetes.

With the right lifestyle choices, diabetes can be managed, and its associated complications can be avoided or postponed. Treatment involves a combination of dietary adjustments, regular physical activity, medication as needed, routine screening, and treatment for potential complications. Caregivers play a vital role in the successful treatment of diabetes as they provide essential support to diabetic patients in managing their condition. Individuals with diabetes have numerous daily responsibilities to keep their condition under control, and having a dedicated caregiver by their side is crucial [5]. Caregivers actively share in the day-to-day responsibilities of their loved ones with diabetes, assisting with tasks such as medication management, blood sugar monitoring, meal planning, and physical activities. In addition to the practical aspects, caregivers also provide invaluable emotional support, understanding the challenges and offering a compassionate presence. By fulfilling these roles, caregivers contribute significantly to the overall well-being and effective management of individuals living with diabetes.

11.1.5 Challenges of caregivers of patients suffering from DM

Caregivers of patients suffering from diabetes face several challenges in their role. According to a study conducted by Niko Dima et al. [6], in Indonesia, caregivers of patients with diabetes face numerous challenges. The researchers examined the relationship between the burden experienced by family caregivers and the health outcomes of individuals with diabetes. The study, which involved 327 participants from the same city, revealed a negative correlation between caregiver burden and the improvement of health status in individuals with diabetes. In other words, when caregivers experienced a higher burden, there was a lower level of health improvement observed in the patients with diabetes.

Some other significant challenges are as follows. (i) Emotional and mental stress: Caring for someone with diabetes can be emotionally and mentally demanding. Caregivers may experience worry, anxiety, and stress related to the patient's health, managing medications, blood sugar fluctuations, and potential complications. (ii) Complex management tasks: Diabetes management involves various tasks, such as monitoring blood glucose levels, administering medications (including insulin injections), managing diet and meal planning, coordinating doctor's appointments, and ensuring regular exercise. Caregivers may need to learn and assist with these tasks, which can be overwhelming and time-consuming. (iii) Education and skill development: Caregivers may need to acquire knowledge and skills related to diabetes management. They may need to stay updated on the latest treatment options, technologies, and dietary guidelines to provide effective support to the patient.

Technology can play a significant role in reducing the burden on caregivers of patients with diabetes. Here are some ways in which technology can help. (i) Remote monitoring: Remote monitoring devices, such as continuous glucose monitoring (CGM) systems, can provide real-time data on blood glucose levels. Caregivers can remotely access this information, reducing the need for frequent finger pricking and allowing for timely interventions if glucose levels are abnormal. (ii) Mobile apps and health trackers: There are numerous mobile apps and wearable health trackers available that allow caregivers to monitor and track important health metrics such as blood glucose, physical activity, medication adherence, and diet. These tools can provide insights and alerts, making it easier for caregivers to keep track of important data and detect patterns or potential issues. (iii) Medication management: Technology-based solutions like medication reminder apps or smart pillboxes can help caregivers ensure that medications are taken on time. These tools can send reminders, track medication schedules, and even notify caregivers if doses are missed. (iv) Telemedicine and telehealth: Virtual healthcare appointments through telemedicine platforms can reduce the need for frequent in-person visits, making healthcare more accessible and convenient for both patients and caregivers. This can save time and effort while still allowing for effective communication with healthcare professionals. (v) Caregiver support platforms: Online platforms and communities dedicated to caregivers of individuals with diabetes can provide valuable support, information, and resources. Caregivers can connect with others facing similar challenges, share experiences, and access educational materials to enhance their knowledge and skills. (vi) Data sharing and collaboration: Technology facilitates easy sharing of health data between caregivers, patients, and healthcare providers. This enables collaborative care and better coordination among all parties involved, ensuring that everyone is informed and can make informed decisions regarding the patient's care.

The research gap and limitations are bridged by the following objectives of diabetic management system:

- Develop a decision support system that assists diabetes caretakers in monitoring and managing glucose levels, medications, and lifestyle choices.
- Provide personalized recommendations and actionable insights based on individual patient data, including glucose levels, meal preferences, medication history, and physical activities.
- Improve caretakers' ability to make informed decisions by integrating evidence-based guidelines, best practices, and real-time data analysis.
- Enhance communication and collaboration between caretakers, patients, and healthcare providers to ensure coordinated care and better treatment outcomes.
- Empower caretakers with the necessary tools and knowledge to prevent diabetes-related complications, promote self-care, and improve overall patient well-being.

11.2 Materials

The method to develop cyber–physical system (CPS) for diabetic healthcare requires the following tools:

A. **User interface design:** Create an intuitive and user-friendly interface for the app that allows caregivers to input the necessary information and view the suggestions provided. *Software:* Adobe XD, Sketch, Figma, InVision, or any other UI design tool.

B. **Questionnaire module:** Build a module that presents the relevant questions to the user and collects their responses, which includes age, gender, etc. *Life style factors:* Hours of sleep, exercise, and food intake factors-amount of meal per day (Number of meals), amounts of carbs+fat+cholesterol. *Medication:* Metformin, Insulin, and Lipitor. *Software:* Backend frameworks like Django, Ruby on Rails, or Node.js for server-side implementation, and frontend frameworks like React or Angular for the user interface. *Methods:* Design a clear and concise questionnaire, implement form validation, store responses securely, and handle data submission. Further, it is good to consult a clinician to validate whether the questionnaire correlates with DM.

C. **Data storage and retrieval:** Develop a system to store and retrieve user data, including their diabetes status, glucose levels, and meal preferences. *Software:* Relational databases like MySQL, PostgreSQL, or SQLite, or NoSQL databases like MongoDB or Firebase. *Hardware:* Server infrastructure to host the database and handle data storage and retrieval (a cloud-based database management system can also be used).

D. **Decision logic:** Design the decision-making logic based on the collected information to generate appropriate suggestions for the caregivers. *Algorithms:* Rule-based systems, expert systems, or machine learning algorithms such as decision trees, random forests, or neural networks. *Methods:* Define decision rules based on the collected information, implement algorithms to process user inputs, and generate appropriate suggestions or recommendations.

E. **Backend development:** Implement the necessary backend infrastructure to support the app's functionality, including data processing and integration with wearable devices. *Software:* Backend frameworks like Django, Ruby on Rails, Node.js, or Flask. *Hardware:* Server infrastructure to host the backend code. *Methods:* Implement server-side logic to handle data processing, integrate with external APIs or wearable devices, manage user authentication and authorization, and provide APIs for frontend communication.

F. **Machine learning and data analysis:** Incorporate machine learning algorithms or data analysis techniques to improve the accuracy and relevance of the suggestions provided by the app. *Software:* Python libraries like scikit-learn, TensorFlow, or PyTorch for machine learning, and pandas, NumPy, or R for data analysis. *Algorithms:* Various machine learning algorithms, such as regression, classification, clustering, or recommendation algorithms. *Methods:* Train machine learning models on historical data to predict glucose levels or suggest personalized recommendations, perform feature engineering and data preprocessing, evaluate model performance, and iterate on model improvements.
G. **Testing and refinement:** Thoroughly test the app's functionality, user experience, and data accuracy to ensure reliable and accurate recommendations. *Methods:* Conduct functional testing to ensure all features work as intended, perform usability testing with real users to evaluate the user experience, validate data accuracy and reliability, and address any identified issues or bugs.
H. **Continuous improvement:** Plan for future updates and enhancements based on user feedback and emerging research in diabetes management. *Methods:* Gather user feedback through surveys or user interviews, analyze usage analytics, monitor emerging research in diabetes management, and plan for future updates and enhancements based on user needs and technological advancements.

The collection of data to create a model, which needs to gather information from various locations, age groups, genders, and different medication history. In fact, there are several machine learning and deep learning algorithms that can be used to design DSS for diabetes management. Here are some algorithms commonly used in healthcare applications:

A. **Logistic regression:** Logistic regression is a simple yet effective algorithm for binary classification tasks. It can be used, for example, to predict the likelihood of a patient experiencing a hypoglycemic event based on various input features.
B. **Random forest:** Random forest is an ensemble learning algorithm that combines multiple decision trees. It is effective in handling complex feature interactions and can be used for tasks such as predicting glucose levels or identifying high-risk patients.
C. **Support vector machines (SVM):** SVM is a powerful algorithm for both classification and regression tasks. It can be used to predict outcomes such as diabetes diagnosis or insulin dosage based on input features.

D. **Gradient boosting methods:** Algorithms like XGBoost (eXtreme Gradient Boosting) and LightGBM (Light Gradient Boosting Machine) are popular for their ability to handle large-scale datasets and provide accurate predictions. They can be utilized for tasks such as predicting glycemic control or personalized treatment recommendations.
E. **Recurrent neural networks (RNN):** RNNs are well-suited for sequential data analysis, making them useful for time series forecasting tasks such as predicting glucose levels. Long short-term memory (LSTM) and gated recurrent units (GRUs) are commonly used RNN variants.
F. **Convolutional neural networks (CNNs):** CNNs excel in analyzing image data, which can be valuable for tasks like diabetic retinopathy detection or analyzing skin lesions associated with diabetes-related complications.
G. **Deep reinforcement learning (DRL):** DRL algorithms, such as deep Q-networks (DQN), can be used to optimize treatment decisions by learning from patient data and feedback. They can help in designing personalized treatment plans and dosage recommendations.

It is important to note that the choice of algorithm depends on the specific task and available data. The performance of these algorithms can also be enhanced by appropriate feature engineering, preprocessing techniques, and hyperparameter tuning. Additionally, the interpretability of the algorithm outputs should be considered in healthcare decision support systems to ensure transparency and trust in the recommendations provided. Finally, we connect our system to an IoT to share this data to the clinician using tele patient health monitoring.

11.3 Methodology

The proposed system consists of mainly three parts: (a) data collection from different sources and integration, to form a centralized database (CD), (ii) formation of electronic health records (EHRs), and (iii) development of decision support systems (DSSs). The following steps explain the methodology shown in Figure 11.3:

a. Data collection: Gather relevant data such as patient demographics, glucose levels, meal logs, medication history, physical activity, and other pertinent information through mobile apps, wearable devices, or manual input by caretakers.

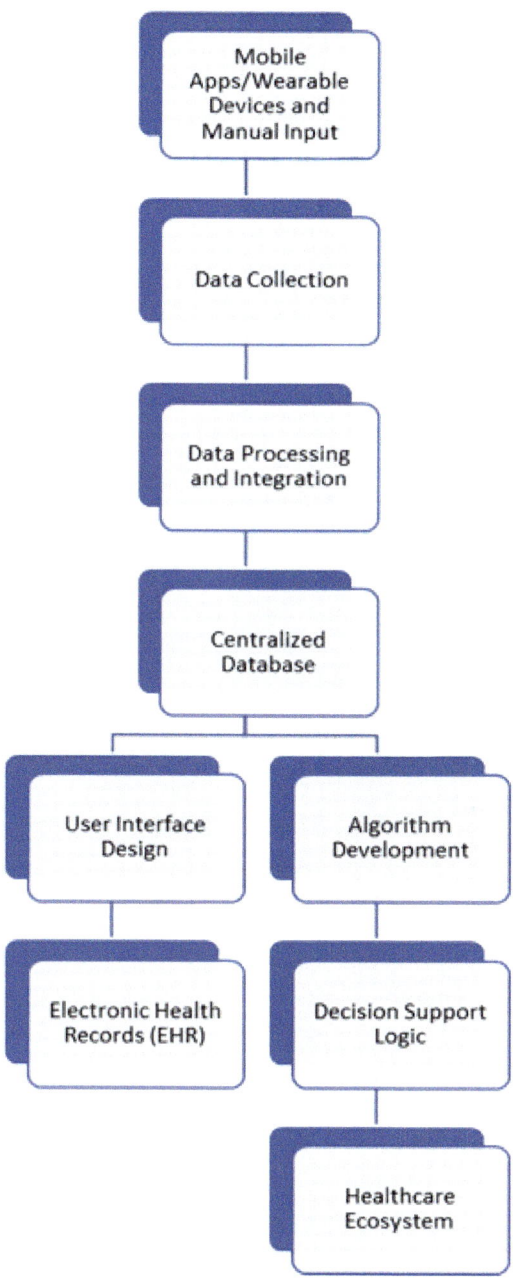

Figure 11.3 Block diagram of the proposed decision support system for diabetic healthcare.

b. Data processing and integration: Clean, preprocess, and integrate the collected data from various sources into a centralized database or platform for analysis and decision-making.
c. Algorithm development: Employ machine learning and data analysis techniques to develop algorithms that can analyze the collected data, identify patterns, and generate personalized recommendations for diabetes management.
d. User interface design: Create an intuitive and user-friendly interface that allows caretakers to input necessary information, view real-time glucose levels, access personalized recommendations, and communicate with healthcare providers.
e. Decision support logic: Implement decision-making logic based on evidence-based guidelines, expert knowledge, and machine learning algorithms to generate recommendations for medication dosages, meal planning, physical activity, and lifestyle modifications.
f. Integration with healthcare ecosystem: Establish interfaces or APIs to integrate the decision support system with electronic health records (EHRs), wearable devices, and other relevant healthcare systems to ensure seamless data exchange and collaboration among caretakers, patients, and healthcare providers.

11.4 Work Plan

The work plan for this proposed chapter is based on the following set of hypotheses:

Hypothesis 1: A methodology along with the associated hardware can be developed to reduce caregiver burden.

Here we are focusing on the development of software and design of associated hardware that can reduce the caregiver burden by providing: (a) an alarming trend on the rise of A1C levels in the last one month or an alarming increase in the glucose level of the patient detected through a wearable device, (b) an unexpected increase in body weight, and (c) changes in the food habits of the patient as self-declared by the patient. This leads to the following sub-hypotheses:

Hypothesis 1.1: An MVP can be developed using a wearable glucose monitoring device informing the caregiver about unexpected and alarming rise in blood glucose levels while correlating it with the changes in self-reported food habits.

A system can be developed informing the caregiver about unexpected and alarming rise in self-reported body weight while correlating it with the changes in self-reported food habits. A methodology that can use all critical variables associated with diabetes and its effects on the disease can be developed for the purpose of synthesizing various decision support systems to improve lifestyle and other important parameters associated with the disease. The system presents a general mathematical framework that can be refined with new discoveries associated with the disease for the purpose of developing decision support systems to improve critical factors in lifestyle of the patient.

Hypothesis 2: An MVP can be developed informing the caregiver about unexpected and alarming rise in self-reported body weight while correlating it with the changes in self-reported food habits.

Hypothesis 2.1: A methodology that can use all critical variables associated with diabetes and its effects on the disease can be developed for the purpose of synthesizing various decision support systems to improve lifestyle and other important parameters associated with the disease.

Hypothesis 2.2: The system presents a general mathematical framework that can be refined with new discoveries associated with the disease for the purpose of developing decision support systems to improve critical factors in lifestyle of the patient.

Value proposition (VP):

VP represented using Figure 11.4 is the reduction in healthcare costs resulting from the use of decision support system will result in reduction in reimbursements for insurance companies by 20%, will result in healthcare costs for patients by at least 10%–50%, and reduce the required skilled workers for hospitals and nursing homes resulting in at least reducing the cost by $500,000.

Mid-term and long-term success indicators:

Mid-term success indicators: i) Development of alpha version of the suggested minimum viable product (MVP), and (ii) submission of patent applications associated with algorithms associated with the MVP.

Long-term success indicators: (i) Successful testing and validation of the MVP, and (ii) submission of manuscripts associated with algorithms associated with the MVP.

Data collection and processing: The wearable device collects glucose readings at regular intervals (e.g., every few minutes) and sends the data to a

254 *Cyber–Physical System for Managing Diabetic Healthcare*

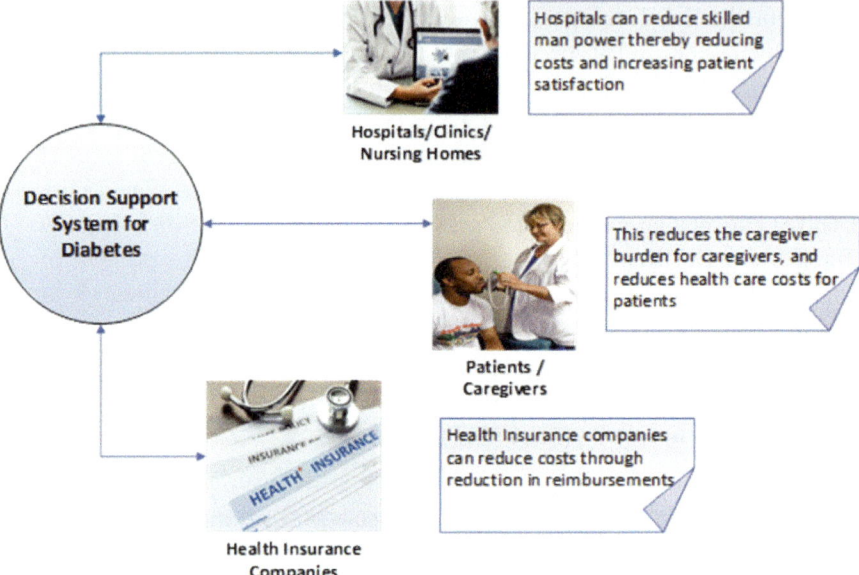

Figure 11.4 Representation of value proposition of proposed work for diabetic health care.

connected device or a mobile app. The app can process the data to identify trends and patterns in the glucose levels.

Threshold setting: The caregiver and user can set specific threshold values for blood glucose levels. These thresholds can be based on individual needs and medical recommendations. For example, a caregiver might want to be alerted if the user's blood glucose levels exceed a certain value, indicating a potential emergency or a need for immediate intervention.

Alert system: When the glucose levels cross the set threshold, the system triggers an alert to notify the caregiver. This alert can be in the form of a notification on a mobile app, an email, a text message, or even a phone call. The alert would provide information about the user's name, current glucose reading, and the severity of the situation (e.g., low, moderate, or high alarm).

Caregiver response: Upon receiving the alert, the caregiver can take appropriate action based on the severity of the situation. This might involve contacting the user, administering medication, providing guidance, or seeking medical assistance if necessary.

Data tracking and analysis: The system can also maintain a history of the user's glucose readings over time. This allows both the caregiver and the user to track patterns, identify trends, and make informed decisions regarding the management of blood glucose levels. Long-term data can provide valuable insights for medical professionals during healthcare appointments.

It is important to note that while a system like this can be highly useful, it should not replace regular medical care and consultation with healthcare professionals. Monitoring devices and systems can provide valuable information, but decisions about treatment and adjustments to medication should always be made in consultation with a healthcare team.

The workflow algorithm of proposed research is represented in Figure 11.4, which is as follows: Step 1 – mobile apps/wearable devices and manual input: This is where relevant data, such as glucose levels, meal logs, physical activity, etc., are collected from patients. It can be through mobile apps, wearable devices, or manually entered by caretakers. Step 2 – data collection: The collected data is gathered and prepared for further processing and analysis. Step 3 – data processing and integration: The data is cleaned, preprocessed, and integrated into a centralized database or platform to ensure consistency and ease of analysis. Step 4 – centralized database: This serves as the central repository for all the integrated data from different sources. Step 5 – algorithm development: Machine learning and data analysis techniques are applied to develop algorithms that can analyze the data, identify patterns, and generate personalized recommendations for diabetes management. Step 6 – user interface design: A user-friendly interface is created to allow caretakers to input necessary information, view real-time glucose levels, access personalized recommendations, and communicate with healthcare providers. Step 7 – decision support logic: This block combines evidence-based guidelines, expert knowledge, and machine learning algorithms to generate recommendations for medication dosages, meal planning, physical activity, and lifestyle modifications. Step 8 – healthcare ecosystem: This represents the broader healthcare system that includes electronic health records (EHRs), other relevant healthcare systems, and healthcare providers.

For the decision support system (DSS) for supporting diabetic healthcare mentioned earlier, the algorithm development process involves designing and implementing the algorithms responsible for analyzing data, generating personalized recommendations, and supporting decision-making in diabetes management. Here are the major steps involved. (i) Data preprocessing: Prior to algorithm development, the collected data needs to be preprocessed to handle missing values, outliers, and noise. Data normalization and standardization may be applied to ensure uniformity and comparability across different features. (ii) Feature extraction: Determine which features (patient demographics, glucose levels, meal logs, medication history, physical activity, etc.) are relevant for the specific diabetes management tasks. Selecting appropriate features can improve the algorithm's efficiency and accuracy. (iii) Algorithm selection: Choose suitable machine learning and data analysis algorithms based on the specific tasks the DSS aims to accomplish. For instance, regression models, classification algorithms, clustering techniques, or time series analysis may be considered, depending on the context. (iv) Training data: Prepare the dataset with labeled examples for supervised learning algorithms or use unlabeled data for unsupervised learning. Divide the data into training and testing sets to evaluate the algorithm's performance. (v) Algorithm training: Train the selected algorithms using the training data. During this step, the algorithms learn from the data to identify patterns and relationships relevant to diabetes management. (vi) Model evaluation: Assess the performance of the trained algorithms using the testing dataset. Common metrics like accuracy, precision, recall, F1-score, or mean squared error can be used to evaluate the models' effectiveness. (vii) Model optimization: Fine-tune the hyperparameters of the algorithms to achieve better performance. Techniques like cross-validation or grid search can help in optimizing the model. (viii) Decision support logic: Implement the decision-making logic using a combination of evidence-based guidelines, expert knowledge, and the outcomes from the trained machine learning models. This logic will be used to generate personalized recommendations for patients. (ix) Integration with user interface: Integrate the algorithm with the user interface of the DSS. The algorithms should be accessible through the UI to receive input data, process it, and display relevant recommendations to caregivers and healthcare providers. (x) Real-time data processing: Design the algorithms to handle real-time data inputs from wearable devices or mobile apps. The DSS should be capable of processing and responding to data in near real time to provide timely support. (xi) Validation and testing: Validate the algorithms' effectiveness in a real-world setting by working with diabetic patients and caregivers. Continuously monitor the system's performance and gather feedback to make further improvements. (xii) Scalability and maintenance: Ensure that the algorithms are scalable to handle a large number of patients and data points. Regularly update and maintain the algorithms to adapt to changing healthcare needs and advancements. (xiii) Compliance and security: Address data privacy and security concerns to ensure that the algorithm complies with healthcare regulations and safeguards patient information.

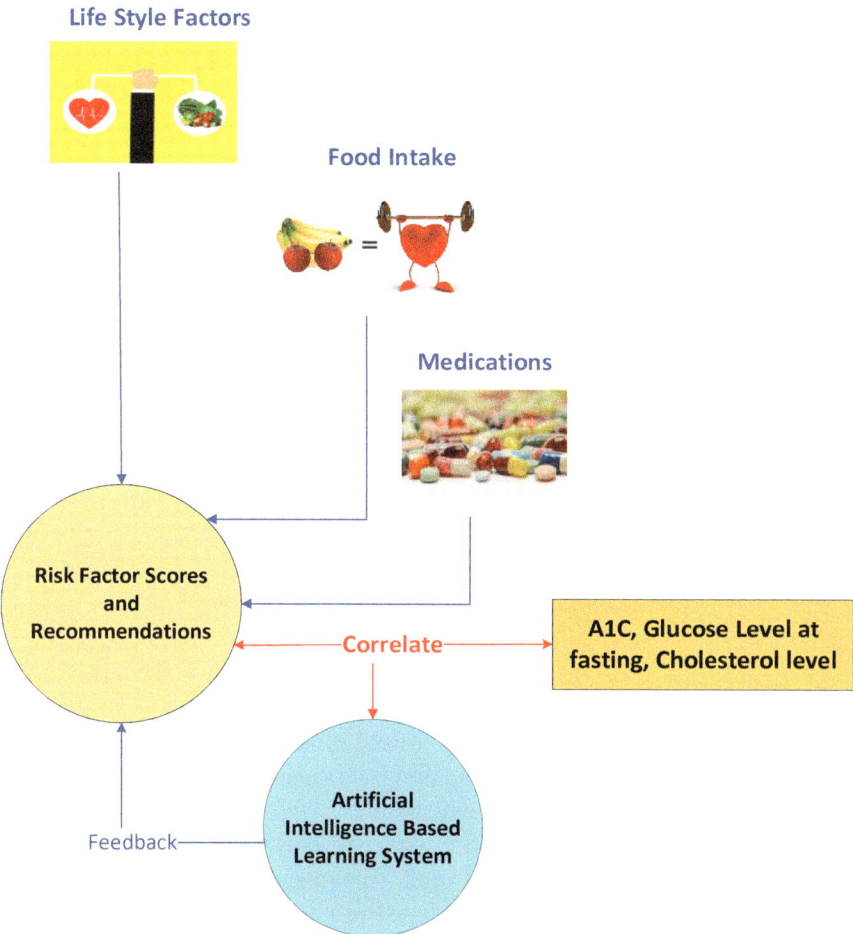

Figure 11.5 Technical details of the diabetic healthcare system.

The above steps are iterative and may require revisions based on the feedback received during the system's implementation and deployment. The goal is to create a reliable and efficient decision support system that aids in effective diabetes management and improves patient outcomes. The technical implementation is illustrated in Figure 11.5.

The broader impacts of this chapter will include: (i) improved diabetes management, leading to better health outcomes and enhanced quality of life for patients, (ii) personalized care that provides treatment plans to individual patient needs, improving treatment effectiveness, (iii) enhanced caretaker confidence in managing diabetes, (iv) better coordination and communication among stakeholders, resulting in a more holistic approach to diabetes care, and (v) empowered caretakers and patients who have access to valuable resources and support for diabetes management. Furthermore, the research outcomes of this chapter can be integrated into education, advancing cyber–physical systems (CPS) education, and disseminating knowledge to healthcare professionals and caretakers. By sharing the knowledge and technological advancements, this work can have a broader impact on diabetes management practices both in India and the United States.

11.4.1 Outcomes of diabetic healthcare system

The outcomes of diabetic's care system are as follows:

i **Personalized recommendations:** Our system leverages individual patient data, including glucose levels, meal preferences, medication history, and physical activity, to generate personalized recommendations. By tailoring the recommendations to each patient's specific needs and circumstances, caretakers can provide more targeted and effective care.

ii **Integration of real-time data:** Our system incorporates real-time data from wearable devices and mobile apps, allowing caretakers to monitor glucose levels and other vital signs in real time. This timely information enables quick decision-making and interventions, leading to better glucose control and proactive management.

iii **Evidence-based decision-making:** Our system integrates evidence-based guidelines, best practices, and research findings in diabetes management. By incorporating the latest research and clinical knowledge, caretakers can make well-informed decisions that are backed by scientific evidence, improving the quality of care provided.

iv **User-friendly interface:** We prioritize the development of an intuitive and user-friendly interface for our system. Caretakers can easily input necessary information, view real-time data, access recommendations, and communicate with healthcare providers. The interface is designed to be accessible to users with varying levels of technological proficiency.

v **Continuous support and education:** Our system goes beyond providing recommendations by offering continuous support and education to caretakers. It can deliver educational resources, reminders for medication administration, and personalized tips for diabetes management.

This ongoing support aims to empower caretakers and enhance their knowledge and skills in diabetes care.

vi Machine learning capabilities: Our system incorporates machine learning algorithms to analyze data patterns, identify trends, and continuously improve the accuracy and relevance of recommendations. These algorithms can adapt to individual patient needs over time, resulting in increasingly personalized and effective care plans. We keep on updating our outcomes based on the development of the research. Also, our machine learning model will keep on updating the new data, and the model weights can be stored in the cloud to make predictions.

A crucial aspect to consider is that the outcomes of our proposed study will yield substantial benefits for the society, given the shared concern over the adverse effects of diabetes around the world. Henceforth, it is critically necessary to develop a system and the associated methodologies and models that can be globally validated for diabetic care.

11.5 Conclusion

Adopting a healthy diet, engaging in regular physical activity, maintaining normal body weight, and abstaining from tobacco use are effective measures to prevent or delay the onset of type-2 diabetes.

With the right lifestyle choices, diabetes can be managed and its associated complications can be avoided or postponed. Treatment involves a combination of dietary adjustments, regular physical activity, medication as needed, and routine screening and treatment for potential complications.

References

[1] Sapra, A., and P. Bhandari. "Diabetes Mellitus. 2021 Sep 18." StatPearls [Internet]. Treasure Island (FL): StatPearls Publishing (2022).
[2] https://www.who.int/india/health-topics/mobile-technology-for-preventing-ncds
[3] https://www.who.int/news-room/fact-sheets/detail/diabetes
[4] https://www.niddk.nih.gov/health-information/health-statistics/diabetes-statistics
[5] https://www.mantachieclinic.org/what-caregivers-need-to-know-about-caring-for-their-diabetes-patient
[6] Kristaningrum, Niko Dima, et al. "Correlation between the burden of family caregivers and health status of people with diabetes mellitus." Journal of public health research 10.2 (2021): jphr-2021.

[7] Wang, Youfa, et al. "A systematic review of application and effectiveness of mHealth interventions for obesity and diabetes treatment and self-management." *Advances in Nutrition* 8.3 (2017): 449-462.
[8] Knox, Emily CL, et al. "Impact of technology-based interventions for children and young people with type 1 diabetes on key diabetes self-management behaviours and prerequisites: a systematic review." *BMC endocrine disorders* 19.1 (2019): 1-14.
[9] Jo, Ara, et al. "Is there a benefit to patients using wearable devices such as Fitbit or health apps on mobiles? A systematic review." *The American journal of medicine* 132.12 (2019): 1394-1400.
[10] Kamei, Tomoko, et al. "The use of wearable devices in chronic disease management to enhance adherence and improve telehealth outcomes: a systematic review and meta-analysis." *Journal of Telemedicine and Telecare* 28.5 (2022): 342-359.
[11] Makroum, Mohammed Amine, et al. "Machine learning and smart devices for diabetes management: Systematic review." *Sensors* 22.5 (2022): 1843.
[12] Peng, Ping, et al. "Effectiveness of wearable activity monitors on metabolic outcomes in patients with type 2 diabetes: a systematic review and meta-analysis." *Endocrine Practice* (2023).

Index

A
Adam optimize 126, 138
AI-based diabetic management 242
Artificial intelligence (AI) 21, 43, 62, 117, 171, 223, 257

C
causality analysis 9
clinical decision 6, 8, 12, 104, 191
CNN-LSTM 205, 217
Coati optimization 176, 181, 189
computer decision support 16
continuous diagnosing 75, 76
cup to disc ratio 145, 148, 151, 155, 157

D
data processing 61, 248, 252, 255, 256
decision support system (DSS) 6, 30, 34, 241, 247, 250, 251, 257
deep learning 7, 109, 114, 125, 172, 195, 201, 204, 206, 210
deep transfer learning 125, 127, 129, 132, 141
diabetes mellitus 2, 4, 63, 108, 170, 223, 227, 242
diabetes prediction 169, 181, 188, 195, 202, 203, 204, 206, 214
diabetes screening 34, 81, 82
diabetes therapy 51
diabetic healthcare 15, 241, 251, 257, 258
diabetic retinopathy 33, 38, 125, 150, 242
digital health 7, 17, 41, 46, 48, 107, 143

E
early detection 7, 24, 72, 113, 147, 196, 223
EfficientNetB3 125, 126, 133
EfficientNetB3-DT 125, 129, 130, 137
expert systems 53, 54, 57, 62, 72, 190, 248

F
feature analysis 172, 175, 177, 178, 179
feature extraction 88, 130, 145, 151, 155, 158, 211, 256

G
gene expression 169, 173, 175, 176, 178, 184
Genomic data 181
Glaucoma 145, 150, 164

H
health informatics 192
healthcare costs 35, 81, 96, 101, 102, 253

I

individual diabetes control program 53, 54, 75, 76
intelligent diagnosis 82, 223, 237
intelligent diagnosis support systems (IDSS) 82
IoT 26, 43, 53, 58, 60, 77, 173

K

K-means clustering 91, 145, 148, 150, 151, 153, 164, 227
KNN 184, 185, 186, 188, 197, 202, 223, 231, 234, 236

L

LightGBM 223, 234, 235, 236, 237

M

machine learning 21, 83, 98, 169, 203, 209, 214, 223, 230, 249
medical data 7, 28, 30, 97, 202
mobile healthcare 45
monitoring sensors 51, 62
multilayer model 169, 172, 175

N

noise filtering 176

O

optic cup 145, 147, 150, 152, 154, 156
optic disc 145, 146, 147, 149, 150, 152, 154

P

patient outcomes 16, 23, 24, 30, 81, 84, 111, 242, 257
Pima Indian diabetes database 202, 204, 223
pre-pre-diabetic 74

S

support systems 7, 30, 34, 44, 81, 250, 253

W

wearable device 7, 25, 26, 44, 61, 98, 115, 244, 245, 253

About the Editors

Usha Desai is presently working as a Professor and Dean (Research & Development) for S.E.A College of Engineering & Technology, Bengaluru. She received her Ph.D. in Biomedical Signal Processing from REVA University, Bengaluru, and M. Tech. and B.Eng. from Visvesvaraya Technological University, Belagavi, Karnataka. She received a DST International Travel Grant to present her research paper at 39th IEEE EMBS International Annual Conference held in South Korea. She has served as Session Chair in reputed IEEE International Conferences. Also, she has presented papers in many reputed conferences and authored more than 40 research publications. She has authored five books in Biomedical Healthcare. She has six patents. She is presently Senior Member of IEEE and Life Member of ISTE.

Biswaranjan Acharya (Senior Member, IEEE) received his M.C.A. degree from IGNOU, New Delhi, India, in 2009, an M.Tech. degree in Computer Science and Engineering from Biju Patnaik University of Technology (BPUT), Rourkela, Odisha, India, in 2012, and a Ph.D. degree in Computer Science from Veer Surendra Sai University of Technology (VSSUT), Burla, Odisha, India, in 2024. He is currently an Assistant Professor in the Department of Computer Engineering-AI and BDA at Marwadi University, Gujarat, India. He is also a recipient of a research fellowship at INTI International University from December 15, 2023, to December 31, 2025. He has more than ten years of academic experience at reputed institutions such as Ravenshaw University, and he has also worked in the software development industry. He has co-authored more than 70 research articles in internationally reputed journals and serves as a reviewer for several peer-reviewed journals. Additionally, he holds more than 50 patents. His research interests include multiprocessor scheduling, data analytics, computer vision, machine learning, and the Internet of Things (IoT). He is currently serving as a secondary IEEE Computer Society representative to the IEEE Nanotechnology Council (NTC) Administrative Committee and as an observer of the IEEE P2851 Standard for Functional Safety Data Format. He is also associated with various educational and research societies, including IACSIT, CSI, IAENG, and ISC.